WE HAVE THE TECHNOLOGY

WE HAVE THE TECHNOLOGY

How Biohackers, Foodies,

Physicians, and Scientists

Are Transforming

Human Perception,

One Sense at a Time

Kara Platoni

BASIC BOOKS
A Member of the Perseus Books Group
New York

Designed by Milenda Lee

A CIP catalog record for this book is available from the Library of Congress.
Library of Congress Control Number: 2015951053
ISBN: 978-0-465-08997-0 (HC)
ISBN: 978-0-465-07375-7 (EB)
10 9 8 7 6 5 4 3 2 1

To my parents, with all my thanks.
And to the wild type and the knockout mouse, with all my hope.

Contents

Introduction   ix

PART ONE The Five Senses

ONE   Taste   3
TWO   Smell   29
THREE   Vision   54
FOUR   Hearing   74
FIVE   Touch   94

PART TWO Metasensory Perception

SIX   Time   117
SEVEN   Pain   138
EIGHT   Emotion   159

PART THREE Hacking Perception

NINE   Virtual Reality   183
TEN   Augmented Reality   203
ELEVEN   New Senses   230

Acknowledgments   255
Notes   257
Index   267

Introduction

IT'S FRIDAY NIGHT, and the denizens of Grindhouse are going to RadioShack.

They cut odd figures in this suburban Pennsylvania shopping center, where the harsh fluorescent lights expose them as basement-dwelling creatures. Wiry, pale, bespectacled, these underground engineers are founding members of the biohacker group Grindhouse Wetware. But Tim Cannon and Shawn Sarver are also minor celebrities around here, partly thanks to many late-night forays to this very electronics outfitter, and partly because of how much of its inventory is lodged in Cannon's arm.

The minute they walk inside the building, the guy at the mobile phone kiosk eagerly waves them over so he can check out their latest project: a thermal-sensing implant buried in Cannon's lower arm. The implant is a silicone-encased slab about the size of a deck of cards, and it lights up like a Christmas tree. You definitely can't buy anything quite like this at the mall.

Grinders are biohackers who build body-modifying, or "body-modding," gear in an effort to enhance human experience. Tonight they need parts for building an implant to put in Sarver's hand. They envision a star-shaped insert that will glow brighter whenever he faces north, turning his hand into a kind of compass. This, they hope, will endow him with a not-quite-innate, but still improved, ability to gauge direction.

"Do we need anything else?" Cannon asks, sifting through drawers of electronic components, bits and pieces to modulate the flow of current. "We've got resistors by the dozen."

"Yeah, we have plenty of resistors," agrees Sarver. They call to each other across the aisles as they browse for parts to add to their stash of lab supplies. Jumper wires? They can always use more. Circuit boards? Sure. Piezo transducer? *"Excellent,"* says Sarver.

Once the grinders have bagged their wares, they stop at a diner where they onboard many fluid ounces of the nerd fuel of choice, Mountain Dew, and get ready to head down to their basement. There, they'll see if they can Frankenstein themselves a new sensory experience.

Grindhouse members are, to put it mildly, underwhelmed by the perceptual apparatus that comes with the standard-issue human body. They see the relative dearth of sensory ports open to us—only five: taste, smell, vision, hearing, touch—as a problem to solve. And even these five have their limits. Why can't we catch up to our colleagues in the animal kingdom and sense the polarization, or directional patterns, of sunlight, the way some bats, birds, and bugs do? Or feel electricity like sharks can? Or pick up ultraviolet wavelengths like the lowly mantis shrimp?

Grinders are the electrical engineering arm of an exploratory community of citizen scientists more broadly known as biohackers. Although computer hackers have a bad reputation for malice or sabotage, biohackers use "hack" in the positive sense of a helpful trick or fix, and I'm going to follow their lead. Biohackers are interested in the organic world, rather than the silicon one of computers. Thanks to the easy availability of genetic engineering information on the Internet and the falling price of lab technology, some biohackers are tinkering with the DNA of plants and bacteria, a practice known as DIYbio, or "do-it-yourself biology." Others try to upgrade their bodies with specialized diets, nutritional supplements, and wearable biometric gadgets that help them track and optimize their sleep, workouts, energy levels, or brain fitness. And grinders: They hack themselves. They are part of a body-modding community that goes well beyond decorative tattoos and piercings. The most ambitious among them are trying to outfit themselves with new machineries of sensory perception.

But whatever their method, biohackers are driven by an urge to create, to enhance, to supersede the ordinary. Nature is amazing, they readily concede, but couldn't it be more so? As Cannon puts it midway through his third Mountain Dew: "Why *not* mess with the body?"

The grinders' explorations are fueled by a steadily boiling, if perfectly affable, impatience: impatience with the weaknesses and limitations of the basic human unit; with the slow pace of evolution; with the unwillingness of major research corporations to develop—and sell—the kind of sci-fi sensory apparatus that biohackers desire to keep things moving along. Their inspiration—and the grinder name—hails from Warren Ellis' *Doktor Sleepless* graphic novels. The series, which began in 2007, portrays a hardscrabble underground populated by body modders: Shrieky Girls who share a sense of touch through networked devices buried inside fake teeth and fingernails. People who instant message each other via their contact lenses. Couples with palm implants that let them feel their beloved's heart beating in their hand.

Doktor Sleepless is a mad scientist (of course) who, through late-night radio, exhorts the grinders to invent a future no one else will. "You can rebuild your own fucking bodies at home with stuff you bought from the hardware store," he urges them. "You're grinders. While you wait for the real future you think you're owed, you fuck around with your bodies like they were virtual-world avatars. You add things to them. You make them better. You treat them like characters to be improved and you grind them." Grindhouse's mantra is also borrowed from the series: *"Where's my fucking jet pack?"* It's a cri de coeur born of out of frustration with a future that, so far, looks a lot like the past.

And so they're trying to hurry up a better future for perception. Grindhouse denizens have implanted magnets in their fingertips in a bid to sense electromagnetic fields. They developed a device that works with these magnets to create a kind of sonar that gauges the distance of nearby objects. The lump in Cannon's arm is a first stab at reading out internal health metrics; if it works, this implant will monitor his temperature. Up next on the Grindhouse wish list is an in-hand compass, a voluntary mutation for anyone who's ever envied a homing pigeon.

Their basic gambit is something like this: Open up your flesh, install a device, and see if it can talk to your nervous system. If it can, you've broadened your sensory world without waiting for slow, clunky evolution to do it for you.

The Grindhouse guys are attempting to leapfrog evolution in their own distinctively homebrew fashion, equipped with the cheapest gear imaginable. But they are also part of a much larger effort to explore one of the most fundamental mysteries of human experience: *perception.* These explorers hail from a variety of scientific disciplines, and nearly all of them require a great deal more credentialing and safety gear than Grindhouse does. They're academics, entrepreneurs, doctors, and engineers. But they're all gripped by the same questions: How much do we know about what happens in our minds as we interface with the outside world? Could we perceive more than we do now? Even if there are hard limits on the brain's perceptual powers, can we work within them to enhance or alter the ways we sense the world?

The world of sensory science is wide and deep. Thousands of people are pursuing similar questions, even if their pursuits are guided by competing theories and motivations. Some are working to restore what is considered "normal" functionality to those without it—whether that means helping the blind see, the deaf hear, or the paralyzed touch. Others want to push past these concepts of normal; they are looking for ways to change or augment sensation through new therapies or wearable devices. And some simply want to know more about how the sensory system—receptors, nerves, and brain— works together to make the world feel *real.*

I've been a reporter for nearly two decades, a science reporter for most of that. But when I began this project, sensory science was, to me, mostly new terrain. I'd previously covered the work of only six of the more than 100 people we'll meet in this book. Still, I found the vastness of the seemingly straightforward question the field confronts irresistible. *Could reality be bigger?*

So I took a year off from my teaching gig at UC Berkeley's journalism school and followed the first rule of our craft: *Get out there.* I sofa-surfed my way through four countries and eight states by posting my itinerary on Facebook, staying with friends, relatives, and fellow journalists who generously opened their homes to a disheveled roving reporter. I haunted the labs, offices, and operating rooms of anyone in the sensory science world who would let me get in on an experiment or a demo. I wore through four tape recorders, 37 notebooks, three rental cars, countless batteries, and one couch, the armrest of which basically eroded under my sneakered feet as I typed and typed and typed up my interview notes. I met neuroscientists, engineers, psychologists, geneticists, surgeons, piercers, transhumanists, futurists, ethicists, designers, entrepreneurs, soldiers, chefs, picklers, and perfumers. I didn't have a grand theory to prove or a particular endpoint in sight; my plan was just to listen and to observe the people who occupy this world. But after a while I began to feel the logic and shape of the terrain. Specific themes kept bubbling up, refrains followed me from interview to interview, and once-disparate-seeming ideas began to click together, usually when I was sequestered in someone's lab, midchew or midsniff or stumbling around in the darkness wearing some kind of weird helmet.

By far the most important thing I learned is this: There is no single, universal experience of "reality," no objective portrait of the world we collectively share. There is only *perception*: what *seems* real to *you*. A percept is the technical name for a mental impression, a sensation, an experience. But a percept isn't reality, any more than an image in a mirror is reality. It's a reflection of the thing, not the thing itself. And as we all know, a reflection can be warped.

That's because the brain is actually a lonely device, an electrochemical jelly cocooned within your cranium, with few ways to interact directly with what's outside. The senses are its go-betweens with the external world, and the information they pass on is always mediated. You can think of the sensory side of the nervous system as its input channels. Scientists call the neurons along this path *afferent*; they carry information to the brain. The system also has output channels, *efferent* neurons, which carry instructions away from the central nervous system—the spinal column and the brain. These outbound channels are the system's motor side, controlling reaction and movement.

Nerves in the sensory tissues—the tongue, nose, eyes, ears, and skin—belong to what's called the peripheral nervous system. This is where receptors, or sensory nerve endings, on or near the body's surface detect chemicals and environmental energies like light or sound waves or pressure. They kick off the process of transducing, or translating, them into brain-ready electrical signals, which the nerves carry. These signals are then collected and integrated by the spinal column and the brain. The brain, specifically, is where these ethereal impulses become, to you, tastes, scents, images, noises, and textures. It's here in the darkened cinema of your skull that the story of your life unspools.

Which isn't to say that this story is necessarily *true*. The brain reads only electrical impulses, and is completely indifferent to their source, which is why we can have perceptual experiences that feel perfectly real even if they aren't. Electrically stimulating the occipital lobe, which processes vision, can create the illusion of a flash of light. Amputees can feel tingling in limbs they have lost. You can taste decadent cakes in your dreams and wake up chewing nothing.

It isn't the *whole* story, either. Your senses take in an unimaginably vast amount of information, much more than you can use. To avoid information overload and maintain a coherent plotline, compression and editing are vital. As we'll see in just a bit, the brain's neural circuitry is continually making executive decisions about how to allocate attention and categorize experiences without your conscious say-so. It *must*. You'd never get through this book if you had to give a thumbs up to each of the many neurons devoted to figuring out, say, where the black letters on this page end and where the white space behind them begins. Time would seem hopelessly jumbled if your brain didn't retroactively edit to make sure that sound and picture always sync, even though hearing is actually faster than sight. Everything would be gibberish if your brain didn't know how to sort noises into words, light and shadow into shapes, and tastes and smells into recognizable categories.

The fact that everyone's machinery filters the incoming world a little differently, sometimes even picking up on data points that elude others, is why there's no one reality. Some of this variation is strictly genetic; there's no question that genes set the limits of our perceptual worlds. Michael Tordoff is a researcher with Philadelphia's Monell Chemical Senses Center; we'll meet him properly in Chapter 1, as one of the leaders in the hunt for a sixth taste, a percept beyond the known five: sweet, salty, sour, bitter, and umami. One day as we were eating lunch, he blew my mind with the news that cats can't taste sweetness. "If you give them a bowl of sugar and a bowl of water, they'll treat them as if they were both water," he said. Like people and other mammals, cats have the gene for a sweet receptor, but theirs is mutated so it can't make a functional receptor. This makes sense in evolutionary terms, Tordoff said, because why would a carnivore need to taste sugar? Sea lions, he added, have an even more limited taste world. "They don't naturally chew their food—they swallow it straight away," he pointed out. "What's the point of having a taste for that?

Even within a species, Tordoff continued, researchers find a good deal of variation. Think about color blindness—about 8 percent of Caucasian men have red–green vision defects, and have trouble seeing or telling apart different shades. Or consider the gene that controls the bitter receptor that affects one's ability to taste a chemical called phenylthiocarbamide (better known as PTC). About 70 percent of people carry at least one variant of the gene that makes them sensitive in some degree to PTC, but others can't taste it at all. Research has suggested this genetic difference might be responsible for people's divergent reactions to the tastes of tobacco, tea, and astringent vegetables like cabbage and broccoli, which contain similar compounds. So while you may find broccoli to be delicious and green, your neighbor might experience it as bitter and . . . well, *ungreen*. As Tordoff put it, "Each animal lives in its own sensory world, and we live in our own sensory world, too."

But your brain's ability to filter—to know which bits of information from life's perpetual data storm to attend to and which to ignore—is a product not only of nature, but of culture, too. Attention can be modified, and that's not purely a function of modern technology. We've been doing it all along. We'll see this in what follows as I move back and forth between what I'm calling the "soft biohacking" of social and cultural forces and the "hard biohacking" of technology. By soft biohacking, I mean the ways we unconsciously learn to pay attention to key sensory information about other people and our surroundings. We experience these soft biohacking forces passively, absorbing them over a lifetime. Among them are language, culture, and routine formative experiences, like the foods we eat, the names we learn for ordinary things, and how the people around us behave and reinforce our own behavior. These teach us what is salient and how to categorize, name, and recall our sensory experiences. Our past experiences frame what we expect the sensory world to be like in the future—and, in so doing, they guide our attention, causing us to dwell on some stimuli while ignoring the rest.

We'll see soft biohacking at work in Chapter 1, as we set out to keep pace with the forerunners in the search for a sixth taste. One of the most confounding puzzles for taste researchers is the word problem of trying to describe a sixth percept that is distinct from the other five. How are you going to find a new taste, if you don't have the language—and therefore the mental construct—for what you're looking for? To be even more blunt: Do you need a word in order to grok a percept? Or do you need that mental concept before you can invent the word?

Words direct our attention to already-established concepts, and a lack of words prevents us from discerning new ones, or at least makes trying to isolate a sixth taste more confusing. As Nicole Garneau, a geneticist who is leading the world's biggest public search for people who can taste fat, puts it, the one thing she can tell you about naming fat taste is that you shouldn't call it *fat taste*. "Because people think of bacon, right?" she said with a shrug before her team of citizen scientists plunged me into a battery of tests to see if I could, in fact, taste fat. (I'll say this now: Pure fat is *nothing* like bacon.)

Next, we'll travel to France, the country that, thanks to Marcel Proust's *In Search of Lost Time*, immortalized the connection between scent and memory. But instead, we'll learn about the link between smell and *forgetting*. Losing the ability to differentiate between scents is one of the first clinical symptoms of Alzheimer's and other diseases of memory. We'll spend time in the Atelier Olfactif with a group of cosmetics industry volunteers who are using odors to help the cognitively impaired recall memories and communicate with loved ones. And here we'll see the soft biohacking of culture at work. Where you grew up changes what you think you smell, because everything about your life experiences—familiar foods, common household products, the kinds of plants that grow nearby—dictates your connections between word and odor. That's why when Alienor Massenet, the atelier's lead perfumer, hands me sample after sample, I repeatedly default to the cultural associations of my California upbringing instead of the associations her French charges would have. To me, lilac is the scent of soap, not a flower. When she gives me the lavender-scented swatch, I think of warm hillsides and it leads me instead to "pine." ("This smell is *so French*," she says forgivingly.) But there are some scent memories she and I share; we both immediately know the smell of the ocean. And when it comes to helping Alzheimer's patients communicate, correctly identifying the scents doesn't matter—only the memories they provoke.

In Montréal, Palo Alto, and Washington, D.C., we'll see how soft biohacking works particularly effectively in the realm of emotion. Our home culture teaches us how to interpret the physical and mental states associated with our feelings, and even how to read the emotions of others. This makes sense, clinical psychologist Andrew Ryder said as we watched students running experiments in his Montréal lab, because in a world of infinite emotional and social data, you want to pay attention to the most culturally significant signals, the ones most meaningful to you and the people around you. "A complex and ambiguous world is forever throwing up new information," he said. "We only want to spend energy on the information that might matter to us."

In Los Angeles and the San Francisco Bay Area, we'll learn how research on pain performed inside the fMRI scanner (by professionals) and inside cocktail bars and taverns (by me) is giving us surprising insights into how we perceive internal states. We tend to think of physical and emotional pain as separate entities: wounds to the flesh versus wounds to the spirit. But social psychologist Naomi Eisenberger, who studies pain as it relates to love and rejection, thinks they are in fact both wounds to the same region of the brain, the part that processes threat. Her jumping-off point is language, how we use identical words to describe social and physical pain—ache, break, and anything else that sounds at home in a country song—despite the fact that we think of them as entirely different kinds of experiences.

"I think there is some prejudice around emotional pain sometimes," Eisenberger said one day in her lab at UC Los Angeles. "Physical pain is totally understandable. Like, of course you are going to feel hurt if you break a leg! Feeling social pain—somehow people often say, 'Get over it' or 'Deal with it. It's just in your head.' And so I think sometimes people feel very validated to hear that some of the same neural regions respond to both. That suggests that we should be taking both physical and social pain seriously." That idea is opening up some surprising new questions, such as, Can you soothe a broken heart with Tylenol, or assuage physical pain by holding a loved one's hand? We have long alleviated the pain of bodily injury with ice packs and aspirin; perhaps now there's a new way to treat the pain of social injury, one of life's most unpleasant—but fundamental—perceptual experiences.

We'll also meet researchers who deal in what I'm calling the "hard biohacking" of technology. I'll be focusing here on devices that people deliberately wear, carry, or implant to alter perception, and this gear is much less passive than social forces in shaping sensory experience. If using technology to manipulate what happens inside your head seems like a futuristic proposition, just consider one of mankind's earliest perception-shaping devices: the timepiece. In Chapter 6 on time, we'll see that this perception is a blending of neural, social, and mechanical forces, one that seems to come from both within and outside the body. In a London museum and at a government laboratory in Colorado, we'll meet the keepers of some very special clocks, one of which was designed to standardize our perception of time, the other to alter it.

And just as the timepiece migrated from rather large, external edifices (think of sundials and clock towers) and then to the tabletop, the wrist, or the pocket, other instruments of sensory perception are scaling down to human size, whether as wearables or, more radically, as implants. Technology is "literally moving towards us like a slow Doctor Who villain," Rob Spence told me one day from his Toronto home. "It's moving into our bodies."

And Spence—better known as Eyeborg—should know. He wears a camera in his right eye socket. We'll learn more about it in Chapter 10, when we visit the explorers of augmented reality, who are trying to enhance human perception through wearable computing devices, commingling human and machine.

Perception-shaping devices are becoming profoundly embedded in human life, partly because they can be worn continuously, partly because they are becoming more deeply integrated with the body. Many of the devices on the commercial market today are wearable, meant to sit lightly on the wrist or over the eye. But other new technologies, currently reserved for those with medical needs, are implanted *in* the body. And their next generation is headed for the brain. So, too, are the scientists exploring the frontiers of perception. In Chapters 3, 4, and 5—on vision, hearing, and touch, all in the first part of this book—we'll trace one very special story arc in modern neuroscience, the ongoing quest to parse the electric language of the brain. This is the study of the twin processes that neuroscientists, in a nod to the analogue world of print, call "writing in" and "reading out." Writing in means moving information to the brain; reading out means interpreting instructions from it.

Our senses are always writing in, taking data from the world and converting it into electrical signals the brain understands. Photons hit the light receptors in your retina, starting an electric relay of signals your brain interprets as an image. Or chemicals lock into the receptors on your tongue, so that your brain registers the resulting electric memo as, say, the taste of sugar. Many of the first generation of write-in devices were built to restore sensory function for people with medical conditions. So we'll spend time with Dean Lloyd, the man with the bionic eye. After years of blindness caused by retinitis pigmentosa, Lloyd became one of the first people to receive a retinal implant, which writes electrical impulses to his retina that his brain can interpret as visual cues. "This is not standard sight that we have," Lloyd cautions, referring to his fellow clinical trial participants. But it is sight nonetheless, and we'll follow along to see how the world looks to Lloyd.

Reading out is the bookend process to writing in; it means translating backward from the brain signal to the sensory experience. For example, if someone shows you a picture or plays you an audio recording, can the pattern of your brain activity be reverse engineered to re-create the original stimulus? Reading out is even tougher than writing in, and before we visit the labs of the people who are attempting it, it's worth pointing out that coming this far has required not only an enormous leap in the ability to translate the brain's language, but calling in the efforts of disparate branches of the sciences. In the early days of sensory science, research was limited to the periphery—the body's outer surfaces, sensory organs and their nerve endings. For example, you might stimulate the taste buds, retinal cells, or skin to learn what the organism would do in response. This was largely the turf of psychologists and physiologists, who correlated stimulus with behavior to understand what was happening later in the nervous system chain. "It's easy to work on the outside, you know," Tordoff, a psychologist, said affably during our lunch. But it's much harder to look inside, he added. "If you can't work out what's going on on the tongue—where it's easy to get to it—how can you work out what's going on in the middle of billions and billions of neurons, all connecting to each other?"

But despite its reputation, the brain is not an unknowable "black box." It is just *very* complicated, and heavily protected by the body and the immune system. Because of the physical and ethical difficulties posed by experimenting on living people's brains, until fairly recently, much of what we knew came from studying other animals. But over the past two decades, a handful of important new technologies brought the collective insights of biochemistry, neuroscience, and genetics to bear on the study of perception. The Human Genome Project unlocked the world of genes and receptors, unveiling the links between DNA and sensory function. Neuroimaging, particularly the fMRI (functional magnetic resonance imaging) scanner, allowed researchers to intricately chart the brain's electrical activity and better correlate stimulus and response. A new generation of multielectrode brain implants has allowed the workings of a living brain to be recorded with striking fidelity.

To learn more about this, we'll stop by UC Berkeley to observe an fMRI experiment in stimulus reconstruction, or reading out brain activity to recreate the original sensory experience. In this case, the subject in the scanner is listening to audio podcasts, and the team eavesdropping on his brain is hoping to decode what he heard. Along with colleagues at other labs, they want to build a model of human hearing so precise that it could be used to read out internal speech—the voice in your head. This ability to translate consciously verbalized words—but not yet much more abstract kinds of thoughts—might help patients who cannot communicate out loud because they are incapacitated by strokes or neurodegenerative diseases.

The final scene in this write-in/read-out triptych will be in the operating room, where surgeon Sherry Wren is at work using robotic arms. These represent a step toward the development of artificial limbs that can not only move with the agility of human hands, but, perhaps, one day experience their same delicacy of touch. The researchers trying to help Wren teleoperate hope to translate this work to prosthetics that could be worn by paralyzed people and controlled by their minds. Developing limbs that can convey inbound sensory feedback from the world—the weight of objects, the force of collisions, the temperature of bodies—while obeying outbound commands from the brain would be the ultimate convergence of writing in and reading out. It's a ballet between organism and machine that must be synchronized perfectly in order not to break the illusion of real touch in real time. This seamless fluidity is the end goal, said Krishna Shenoy, the Stanford University neuroprosthetics expert whose lab we'll visit. Researchers want to become so fluent in this synaptic language that they can "just have a conversation with the brain."

Most write-in and read-out technologies are highly experimental and invasive, largely the purview of research universities and hospitals. But you don't have to undergo surgery to hack perception. In the final stretch of the book, we'll meet the people behind some of the homemade and consumer-grade devices that let you warp your senses for exploration—and fun.

We'll start with the devices that are the most external to the body, and end with the ones inching closest toward it. First, we'll spend time in virtual reality, an entirely external technology that can make use of not just helmets or goggles, but entire high-tech rooms equipped with surround sound, vibrating floors, and even pumped-in odors, to create scenarios that seem so realistic they trick the brain into altering behavior. We'll strap on the goggles at a military base where researchers are testing whether "preliving" awful combat scenarios before deployment can help soldiers be more resistant to post-traumatic stress disorder. Then we'll wear the helmet at a Stanford lab where subjects are asked to virtually inhabit strange bodies and perform bizarre tasks—flying, punching balloons, giving themselves a scrub-down—to see if it prompts them into better social or environmental habits. (Not to give too much away, because part of the perceptual magic behind virtual reality experiments is that we can't see the trick coming, but let's just say that after my time on a virtual farm I'm permanently off hamburger.)

Next, we'll enter the world of augmented reality wearables: glasses, watches, and other gadgets you put on your body—but not *inside* it—to enhance sensory perception. This is a world full of designers who ask out-of-the-box questions, like *Why do we not have night vision, or eyes with automatic zoom? Could we taste a flavor impossible in nature? How can you send an invisible hug?* This is where we'll find Eyeborg, as well as a slew of entrepreneurs hoping to deliver a new generation of augmented reality gear to the mass market. We'll meet the engineers behind iOptik, an augmented reality system that you wear extremely close to the body—a contact lens goes right on your eyeball—and Adrian David Cheok, the pervasive computing professor whose lab is investigating using rings, phone apps, and even fake lips to convey touch, taste, and smell experiences at a distance. To Cheok, the key difference between virtual and augmented reality is that wearing lightweight technology, rather than gazing into helmet-mounted computer screens or being stuck in specialized rooms, means you can move and interact with others normally, making your "mixed reality" experiences more engaging and natural. "Rather than the human being put into a virtual reality system," Cheok said, "now we do the opposite: We bring the virtual world onto our bodies."

This book ends in a basement.

It has to.

The basement—and its Silicon Valley counterpart, the garage—is the birthplace of technical innovation, where our species prototypes its fever dreams, where we discover and hack and build our test creatures. This is where people are hoping to fast-forward evolution, to skip the millennia of random mutation it would take for nature to offer up a new sensory port. So we'll reconnect with the Grindhouse crew and others who are trying to expand perception by modding themselves. Their quest: To write in a class of environmental information humans can't otherwise perceive, primarily, electromagnetism, since their experimentation generally begins by implanting a magnet. My quest, as a reporter, is to figure out if there's a neuroscientific explanation for what happens when they try.

One last word about what it means to hack perception. Soft biohacking through social forces is powerful because it is pervasive, stealthy, and hard to control. We are usually unaware it is even happening and only glimpse its influence when we are jarred by a cultural shift, like traveling to a place where the language or social cues are different. But because these attention habits are learned, they can be relearned. Consider again the language problem that dogs the search for a sixth taste: We don't already have a word or concept for what that taste might be. But people who were in elementary school before the 2000s might remember an era when there were only *four* tastes, not five. As you'll read in Chapter 1, the story of how scientists found the fifth, and how the rest of us learned to perceive it, says a lot about the brain's ability to adjust.

Hacking the brain with technology will be more powerful still. It's not a totally new idea—you can easily consider psychoactive drug use a form of brain hacking, or storytelling, with its power to transport you to imaginary worlds. But until now, these hacks have been short-lived, temporary breaks from the "real world," not efforts to supplant it. As people begin altering perception with devices that interact directly with the sensory system and can be worn continuously, like smart watches and glasses, we are moving into an era in which we can deliberately control perception in a more lasting manner, and perhaps enhance our ordinary selves in extraordinary ways.

For some, like the Grindhouse folks, the prospect of installing perceptual gear of our own choice and invention atop our existing mechanisms is liberating, a way to speed up evolution. Others, like members of the group Stop the Cyborgs, whom we'll meet later, see a risk that we'll create illusions of "reality" that are harder than ever to escape, their influence more difficult to spot, as we entangle our own perceptual apparatus with machineries built and controlled by others. As they point out, technology is never neutral; it is a way of filtering our experience through someone else's design. By putting an artificial scrim between your senses and the world, these technologies might give you superpowers, but they might also limit or mediate your attention and experience, or change your behavior in meaningful ways. And because the influence of these devices is so constant and subtle, you might not notice.

But as we'll discuss with these critics, influencing behavior is human—we already guide each other's thinking with language, with culture, with social interactions. What is new is how closely you can knit sensory perception to a machine, and—should these technologies gain broad acceptance—to an electronic network of other users, making that influence widespread and perhaps even anonymous. We have in some ways been biohackers all along, shaping each other's realities through the mere act of living together. Now, as technology users, we may be able to shape our realities again, this time voluntarily and through glossy consumer gadgets.

This won't be a trivial choice. We are a species on the cusp of tinkering with its own evolution, of driving technology deeper into the flesh, away from the periphery and toward the seat of perception. We are learning to interpret the brain's language, the data flow that turns electrochemical fizz into sensations, experiences, feelings—the very stuff of *being*. And if we can understand this information, we can modify it. To borrow once again from the analogue world of print: As any hack knows, it is powerful to read and write. But it is more powerful still to control the edit.

PART ONE

The Five Senses

ONE

....

Taste

MIKE ARCHER CAREFULLY LAYS OUR SUPPLIES on the lab countertop: A DNA collection kit. A kitchen timer. A bottle of water. A bag of oyster crackers. A green file folder, which he opens to reveal a set of plastic baggies taped neatly inside, each containing a gel wafer the size of a postage stamp. There's also a mysterious device with puffy foam disks on its pincer ends. These are my nose plugs. Archer mimes how to gently clamp my nostrils shut, so I can taste but not smell.

"I want you to do a little sniff test to make sure you don't have any air going through your nose," he says.

I try them on; it's like instantly having the world's worst cold.

"Perfect," Archer says. The experiment is ready to begin.

Behind us, it's a snowy school day morning at the Denver Museum of Nature & Science. Eddies of jubilant kids on field trips, wearing borrowed lab coats and goggles, blow through the adjacent biology exhibit, where they can extract DNA from wheat germ or measure the sugar in breakfast cereals. Every now and then they glance curiously through the enormous plate glass doors that separate this working lab space from the rest of the museum.

Archer, a retired dentist, is a volunteer, part of a small army of "citizen scientists" who power the Genetics of Taste Lab, the formal name for our setup on the other side of the glass. The lab coat he's wearing is mostly for show, something that delights the kids, he says. But the test we are about to do is real; if it works, it will shed some light on one of the biggest mysteries in sensory science. We are going to see if I can taste fat.

Not bacon.

Not cream.

Just *fat*. Or, more specifically, fatty acids. Or, even more specifically, linoleic acid, an omega-6 polyunsaturated fatty acid, which is vital to the human brain and immune system and for that reason is something researchers think our bodies might be able to sense in our food supply. On the wall, which has been painted to look like an enormous lab notebook, there's a diagram of a linoleic acid molecule, its two arms shaped like a ladder bent to the right. If I—and the 1,500 museum visitors taking part in this experiment—can taste fat, we will have helped prove whether or not there are more than five basic tastes. Or, to put it more bluntly, we will have helped show whether there are still unnamed, unexplored dimensions to a sense we thought we already knew.

The five basic, or primary, tastes are listed on an easel propped up on the lab counter: *salty, sweet, sour, bitter,* and *umami,* which is sometimes described as savory. These are considered the building blocks of taste, essential parts which cannot be further subdivided, just as red or blue cannot be fragmented on the color wheel. Chemicals in food bind to receptors that are embedded in bulb-like clusters of cells that are called taste buds, and this information travels along the gustatory nerves to the brain, where it is ultimately interpreted. Some of you may remember a time when there were only four basic tastes; umami didn't officially make the list until 2000, although the concept had already existed for a century in Japan. It was discovered in 1908 by Dr. Kikunae Ikeda, who argued it was a fifth taste associated with the amino acid glutamate. Its acceptance upended the food research world, hinting that the taste universe was bigger than previously thought, casting doubt on the definition of a basic taste itself, and fueling a quest to see if there are more primaries. Much like seventeenth-century astronomers who suspected there were planets beyond the orbit of Jupiter, scientists are now searching for new candidates that would enlarge our known system.

Fat, championed by nutrition scientist Dr. Richard Mattes at Purdue University, is one of the leading contenders. "We think that fatty acids generate a unique sensation," he says, something separate from the other five.

While the taste research community has no formal list of what qualities are required for basic tastes, Mattes has come up with six well-respected rules of thumb. Among them: There must be receptors for the taste stimulus on the tongue. (Two fatty acid receptors, CD36 and GPR120, have been localized to taste cells, and Mattes expects others might be found, too.) That information must travel to the brain along the gustatory nerves rather than the trigeminal nerve, which conveys information about touch—information that in the case of eating is called "mouthfeel." This last requirement is particularly confounding when it comes to fat, which is clearly also a mouthfeel and indeed a reason why creamy, fatty foods are so appealing. "That's one of the big issues—is it really taste or is it just texture?" Mattes says. "Differentiating those is very, very, very, very, very tricky, because to taste something you have to have it physically in contact with your tongue."

In human studies, researchers often try to separate taste and texture by administering the fats as liquids, with mineral oil added to mask lubricity and gums to mask viscosity. Nose plugs are used to block odor, and some testers use blindfolds or red lights to eliminate visual and coloration cues. With rodents, people have tried cutting the gustatory nerves, and found that the animals became less sensitive to fatty acids, suggesting the information is carried on that nerve, not just the trigeminal one, Mattes says.

According to Mattes' guidelines, a basic taste must also serve a biological purpose. Most taste researchers believe the primaries fall into one of two broad camps: attractive and aversive. They generally agree that we are attracted to sweet taste, which signals where the carbohydrate energy is, and umami, which signals the presence of protein, specifically, amino acids. We avoid bitter, which often indicates the presence of toxins. With saltiness, which signals the presence of essential electrolytes, and sourness, which indicates acids like vitamin C, researchers don't always agree, but some think it's probably a matter of concentration and satiety. We eat a little of the foods containing these chemicals when we need them, but they may become aversive at extreme concentrations or when we've had enough. (Mattes points out that people eat all of these things, sometimes simply for pleasure, and that any of them can become repellent if you've had too much.)

A basic taste also prompts a physiological response. Fat usually comes in the form of triglycerides, which contain fatty acids. While free fatty acids are found in some foods, in other cases when their concentration is too low for detection, the body must produce a salivary enzyme called lipase, which chips the fatty acids off the triglyceride. "Mice have plenty of it, and it's been shown very clearly that when you block the activity of this enzyme they don't detect and prefer fat," says Mattes. But there are questions about whether humans produce enough lipase to break down triglycerides, producing enough of these acids to let us taste fat. Mattes has shown that chewing fatty foods like coconut and almonds prompts the release of lipase—enough lipase, he thinks, for us to taste fat. So there's a possible physiological response: Chew the fat, produce more enzymes. (One related hypothesis, says Mattes, is that hard fats like coconut prompt a bigger response than soft ones like olive oil: The harder you chew, the more you salivate, the higher the lipase concentration.)

But the ne plus ultra is whether anyone can actually *sense* the taste, and whether that sensation is distinct from the five we already know. That's where the museum comes in, with its constant influx of visitors of all genetic backgrounds, many of them arriving in family groups (always a boon for genetics researchers). Like me, each participant will swab the inside of their cheek with a giant Q-tip and hand it over for testing. Like me, they'll learn to "swish and swallow" from the water bottle to cleanse the palate between tests, and be directed to make use of the oyster crackers in case they encounter a taste that is exceedingly repellent. And then they'll try some fatty acids.

Archer opens the file folder and extracts the first of the wafers. It's a gelatin square the color of onion skin, and perhaps a bit thicker. If you've ever used a meltaway breath strip, you've seen this technology before. But instead of minty freshness, each strip is infused with a different concentration of linoleic acid.

Archer explains how to do what's called a scaling test. I'll place each strip as far back as I can on my tongue, holding it there for 45 seconds. We'll do three practice strips, and then four more to test whether I can perceive the linoleic acid at different strengths. After each exposure, I'll rate the strength of the sensation. The concentration of each strip will vary, perhaps including placebos with no fat at all. "This is a double-blind study, meaning you don't know what these are and I don't know what these are," Archer says, gesturing toward the file.

All right, nose clips on? "Nose clips on," he says.

The first two practice rounds are easy. One strip, Archer tells me straight out, has no taste—it's just to get me used to the feel of the wafer. It's gummy, and holding it carefully on the tongue while resisting the urge to bite through it conjures some distinctly Communion-related memories. Another strip is laced with one of the five known basic tastes; my job is to identify which. (I guess sweet.) The third round is harder. This wafer has linoleic acid in it; this is the only time in the study I'll know for sure that there's fat to be tasted. The point of this round is to clue my taste buds in to what they're searching for once we launch the true tests. It'll also be my first hint about whether there's something to this fat idea, or whether I've joined a gustatory wild-goose chase.

Archer proffers the gel strip and readies the timer. "Close your eyes. Just kind of concentrate on what you're experiencing, OK?"

At first, there is nothing. Maybe 15 seconds of gelatinous blandness. And then there is a wash of . . . something. My mouth puckers. The first word that springs to mind is *bitter*.

But that's already a basic taste. Try again.

Acid, my brain suggests.

But acid is essentially sour, another basic taste.

My mind flails. How am I ever going to describe a possible sixth sensation without using the terminology of the other five?

And here I have run into the secret second purpose of the museum experiment. We're not just here to see if we can taste fat. We're here to see if we can describe *how fat tastes*. Searching for the sixth taste, it turns out, isn't only a technical problem; it's a word problem.

The lexical challenge hinges on this: How can you perceive something if you don't have a word—and therefore an established concept—for it? In order to recognize a new basic taste, we'll have to train ourselves to discern a separate, distinct quality among the foods we have been eating all along. Taste researchers often compare this conundrum to the idea of isolating a new color. The rainbow's not going to grow, but we might find a new way of categorizing its light, of seeing a particular bandwidth as a discrete, unique entity. There's linguistic precedent for this kind of sensory category sorting: researchers point out that not all cultures break down the light spectrum the same way. Some languages don't have separate words for green and blue, for example. That doesn't mean people who speak those languages can't see both colors. It means that they perceive them as a singular experience, while people raised with a language that distinguishes between them perceive them as separate. And as the German philologist Lazarus Geiger determined in the nineteenth century by studying ancient texts, early cultures tended to develop words for colors in a similar order: black and white first, then red, yellow, green, and finally blue. While at the time he wondered if this might be the result of anatomical evolution, today it can be read as either a linguistic or conceptual change—depending on which side of the debate you're on. And so it may be with taste. Maybe fat's been in the taste rainbow all along; we just don't have the vernacular to call it out.

"It's a total language problem," says Dr. Nicole Garneau, the museum's curator of human health, shaking her head knowingly. "And we stumbled upon that right away." Garneau is an energetic young geneticist with bright blue eyes and a rapid-fire way of speaking who came to the museum after an early career studying yeasts and viruses. The museum wanted to teach genetic concepts in a personalized way, one that would allow visitors to learn about their own bodies and participate in ongoing scientific work. The team focused on taste because the related genes are so individually variable, and zeroed in on the question of a sixth taste because they thought it would excite visitors more than donating their DNA to the umpteenth study on, say, bitter perception. "We wanted to ask questions and do research that is potentially high risk and high reward, because that's what the public is interested in," says Garneau. "They want to be at the cutting edge."

But just putting out a press release advertising the study forced the museum staff to confront the language problem: What were they going to *call* this phenomenon? Their solution, for marketing purposes at least, was "fatty acid taste," a phrase that is precise, if not exactly satisfying. The problem,

Garneau points out, is that "fat taste" doesn't evoke a particular mental image. We have no percept for what fat should taste like, and no descriptive language for it. Our descriptors were built for the first five tastes, and we tend to stick to them even though they are frustratingly limited. "If I give you black coffee and I say, 'Describe this, but don't use the word "bitter,"' or I give you sugar and say, 'Describe this, but don't use the word "sweet,"' it's very difficult to do," Garneau says. "So that's the pickle we're in right now."

The museum's way around this mental roadblock is crowdsourcing. Maybe after free-associating their reactions, the 1,500 participants will collectively describe this thing in a way that will let their fellow humans recognize it. "That's 1,500 brains that are a lot smarter than just one scientist, who is going to use some science-y name that isn't going to be relevant," says Garneau. "We're going to have something simple, that is relatable, that the community can embrace and say, 'That's the sixth taste, and I understand what it means.'"

Archer clicks the kitchen timer. Time to come up with some relatable words. Archer warns me not to evaluate how pleasant the taste is. "I am not interested in good, bad, or gross," he says. "That doesn't tell me anything."

So I say the only word I've got that isn't already a basic taste: varnish.

Archer nods, pushing a paper toward me. Cleaner, I write. Solvent. Not too astringent. Not like, say, Pine Sol. Mellower. *Funkier.*

Next, Archer has me take off the nose clips. The experience most people call taste is more accurately known as flavor, the combined work of the mouth and the nose. When we eat, scent molecules travel from the mouth up the nasal cavity, and the brain processes them in combination. (The scientific term is "retronasal olfaction.") Flavor perception can be dramatically altered by odor, and so the museum wants to test this, too. "Let's take off the clip before you lose that entire thing. What do you think it smells like?" he asks.

The scent is already vanishing, as ephemeral as dream logic. "Just the faintest whiff of shoe polish," I finally say.

Now it's time for the scaling tests. My task is to determine if I can sense that taste again and rate its strength by drawing a mark on a scale line. The next four gel wafers will each contain a different concentration of linoleic acid; I place each in turn on my tongue as Archer mans the stopwatch.

The first tab is weak, the ghost of wood varnish. I mark what would be 1 on a scale of 10.

The second is much stronger, a 7 now, still that lacquer-like tang, although this time when Archer asks me to remove the nose clips, an extra level of funkiness emerges. "There is something old about it," I say. "Stale air. Or suitcase leather or old luggage."

And also: Is my mouth watering? If so, is it because it's almost lunchtime and these wafers are the first food-like objects I've put in my mouth all day? Is it the autosuggestion of the writer who read a whole passel of studies about fat taste before coming to taste fat? Am I salivating because people in Mattes' coconut study did it? *Heck, am I salivating because mice do it?*

I am sure about one thing: Whatever is happening is not fun. On the third try, I nearly become the first person to break into the emergency oyster crackers. Not only is the taste sharper, but the acid feeling in my mouth has progressed to the pit of my stomach, where that distinctive "too much coffee" digestive drumbeat is starting up. After the longest 45 seconds yet, Archer finally stops the timer. Definitely a 9.

"Ugh," I manage. "Brlrlrlrhgghghgh." I try to come up with some more scientifically useful words. Plastic? PVC pipe? Shoes? I've defaulted to metaphor. All my adjectives have failed.

Time for the last trial. I keep an eye on the crackers. But this tab is, I am pretty sure, the placebo. I sense nothing, and my contribution to taste science is officially over. I'll have to wait until 2016 to see how the other 1,499 tasters fared.

For the record, most people don't do much better. When I ask Mattes if he can describe fat, he flails, too. "No! No, I can't! It's awful. Period!" he says. "It sucks the air out of your lungs. It's just terrible. Nauseating. It's rancidity. It's really, really bad cooking oil, the sensation that that gives you." At first, most people default to a familiar basic taste: bitter. "But we don't think bitter is actually the word," says Mattes. "Bitter is just being used as a way of saying it's awful."

There's a reason, Mattes thinks, for why fat should taste unpleasant, despite its texture being enjoyable. We need to eat fats, he says, "so having something that would direct us to it and encourage us to consume it makes sense as much as for any nutrient." But free fatty acids are generated when food spoils, and it's counterproductive to eat food that makes us sick. So, he says, "the role for fat taste is we believe more akin to the role of bitter stimuli—is an aversive stimulus. It's saying, 'This is probably not a wholesome thing to consume.' And so it's to be avoided."

Not everyone finds the linoleic acid quite so unpleasant, or has such a strong reaction. Archer describes it as a musty, mushroom-y, earthy taste. Lab tech Leta Keane, who has been working behind us assembling file folders for future visitors, describes the taste as old, stale oil. "You know like when you leave a fast-food bag in the bottom of your car?" she asks. Garneau says she had distinctly separate taste and smell impressions. The taste, she says, "is very pungent, like a Kalamata olive would be pungent, but without the salt." The smell, she says, is "stale popcorn paper."

In the first six months of experimentation, the most common taste descriptors the museum collected were "bitter," "buttery," and "nothing." For smell, the top four were "nothing/unknown," "paper," "cardboard," and "plastic." But there have been outliers, people who compared the taste to rotten icing, ocean water, dandelions, or gummy bears, and the smell to seaweed, pine nuts, stamp books, or bricks. In April 2015, the researchers released some preliminary results, using data from the first 733 tasters, and found that so far, people were proving that they could indeed taste fatty acids. People in that early pool were accurately rating intensity higher in the wafers that had more linoleic acid. Interestingly, women and girls were more sensitive to it than men and boys, and children more sensitive than adults.

That variability of experience probably has to do with genetics. The CD36 gene is a gnarly one, with 84,090 base pairs. (Compare that with the TAS2R38 gene, one of the many related to bitter taste perception, which has only 1,002 base pairs.) That's a lot of opportunities for small changes, which may affect gene expression—and perhaps much broader changes in receptor function and even eating behavior. One of the theories Mattes' lab has considered is whether sensitivity to fat correlates with body weight: If you are sensitive to it, perhaps you avoid fats and have less fat in your diet. "I don't know if it's right," Mattes says, "but it's interesting."

As a result, the museum is collecting information on participants' eating habits and body types. Between tasting trials, Archer walked me through a series of tasks, including listing how often I eat fatty foods like salad dressing, peanut butter, and eggs and measuring my height, weight, fat mass, and body fat percentage. (The museum's preliminary results from early 2015 found no correlation between body fat and intensity ratings.)

And there's another X factor. When it comes to perception, especially for novel stimuli, people are enormously suggestible. Thanks to Garneau's experience studying yeasts, she has a sideline working with brewers and wine makers—fields where descriptive language for taste is something of a fetish. Wine has an enormously rich and complex language, but one that's notorious for exaggeration and wishful thinking, especially when attempted by amateurs. Garneau recalls attending a tasting at which her husband, just to see what would happen, suggested that the vintage they were drinking had a "gunpowder" quality. Much to her alarm, others at the table said they could taste it, too. Faced with a blank slate, she says, even when someone tosses out an unlikely term like "gunpowder," "you're going to latch onto it. Your brain's going to be like, 'Yeah, that's it! Because I can't come up with the language!'"

She pauses. "And that's using stuff that we *know*," like the flavor of wine. With a potential sixth basic taste, she says, "Now we're talking about a topic that we literally have no language about. So it's compounded. Your brain wants to solve. Your brain *so much* wants to solve this puzzle of what it's tasting. And it doesn't have the language. So whatever anyone throws out there, you're going to be like, 'Yes, that's it!'"

DR. GARY BEAUCHAMP AND DR. MICHAEL TORDOFF could not disagree more about how to unpack this language problem—although they do agree that there is one. Beauchamp is the director of the Monell Chemical Senses Center, an independent research group in Philadelphia that studies taste and smell. It's a sort of Wonka factory for sensory research; on any day you can see people swabbing and spitting their way through mysterious taste tests and mice chowing down on pastel color-coded food pellets. Tordoff is one of the center's leading researchers, and is fielding his own candidate for the sixth taste: calcium. The two scientists, whose offices are just a few floors apart, have landed on the opposite ends of the sixth taste question, and the debate going on between their floors more or less comes down to the word problem: Does perception drive language or does language drive perception?

The answer probably depends on where you look. And where you look has a lot to do with which branch of the sciences you hail from. Since the turn of the twenty-first century, the study of taste has transformed from a mainly psychological science into an increasingly biochemical one, with less focus on how the brain parses the external world and more focus on probing the tongue to identify its receptors. Beauchamp's background is in comparative and behavioral biology; Tordoff's is in psychology. Neither is a biochemist, and both of them call the change in direction nothing less than a revolution. "You can't describe how big the shift has been," says Tordoff. "It's incredible."

Early taste studies focused largely on behavior, often gleaned from observing the feeding habits of mice and men. By cataloguing food-related actions—did the subjects eat more or less? Did they spit it out?—you could draw conclusions about what was happening inside the brain. Appetite, avoidance, eating patterns, preferences—all of these offered clues as to what the internal circuitry demanded, although they didn't necessarily reveal the design of that circuitry.

Scientists have long wanted to get at the biochemical link between what we want to eat (calories!), the specific chemical we sense on the tongue (sugar!), and the perception in the brain (sweet!). One of the center's first goals in the 1970s, Beauchamp points out, was to identify a sweet receptor, but it just couldn't be done with the tools of the time. Then, around the turn of the century, genetic technology hit a boom time. The Human Genome Project finished cracking our DNA, creating an easy way to look up sequences that might be taste receptors. The improvement of the polymerase chain reaction method, or PCR, allowed scientists to very quickly analyze those sequences to see if differences in genes led to differences in proteins that affect taste-sensing capabilities. Transgenic lab animals, with the genes scientists wanted to study suppressed, expressed or swapped, became readily available. Being able to manipulate these animals' DNA made it faster and easier for scientists to test the relationship between gene, receptor, and tasting ability.

And results started to flood in. In 1999, scientists first characterized a new class of G protein–coupled receptors they believed were linked to taste perception in mammals. The family of bitter receptors was first described in 2000; there are probably at least 25 of them, researchers now believe, which help us sense thousands of different compounds. Then in 2001, several groups, including one from Monell, identified a sweet receptor. In 2000 and 2002, scientists found receptors for umami, or more specifically for glutamate. Japanese scientists had been pressing their international colleagues to accept umami as a basic taste for some time. But while the word was quite well understood in Japan, it meant nothing abroad. To Western scientists, savoriness seemed a vague concept. It took the discovery of corresponding receptors on the tongue to persuade them that umami was for real. For the field, this was a symbolic moment, a sign that research had shifted its focus from brain to tongue. It emphasized the opposite end of the tasting process—instead of focusing on the "Aha!" moment of perception and identification, now the focus was on the moment molecule met flesh, when chemical locked into receptor.

After umami, people asked: If there are five basic tastes, why not six? Why might there not be thousands, just as there are countless odors? After all, smell is taste's fellow chemical sense. We don't use primaries to describe scents, nor do we have an accurate tally of the nose's hundreds of odor receptors.

The acceptance of umami bolstered the arguments of those who'd long been questioning the usefulness of primaries. In 1996, psychologist Jeannine Delwiche, then at Cornell University, wrote one of the most famous salvos against the idea, a journal article with the stark title "Are There 'Basic' Tastes?" She argued that the then-four primaries were ill defined and limiting. Asking research subjects to sort their perceptions into four existing categories had created a "self-fulfilling prophecy," she wrote, which crowded out the possibility of nonstandard answers. While she didn't have an alternative system to propose, Delwiche wrote, "It is wiser to accept the multiplicity of taste sensation without relying on a flawed organizational structure."

Today, most researchers agree that the final count is more likely to be something like "close to five" than "thousands." But, Tordoff asks, where do you draw the line? Remember those 20-plus bitter receptors, each sensitive to different compounds? "If you take the molecular biology approach that everything that has a receptor is actually a basic taste," he asks, "are those 20 different bitter tastes? Is each one a *separate* basic taste?"

Where you fall on the basic taste question is "almost philosophical," says Beauchamp—and that has to do with your belief about which part of the process matters most. Beauchamp's personal view is that basic tastes have nothing to do with finding receptors on the tongue; to him what is important is what happens in the mind. "The issue of basic taste is a psychological phenomenon," he says. "It's a perceptual phenomenon, not a mechanistic phenomenon." The key thing a basic taste must have, he argues, is a percept—the mental image, the easily recognizable feeling that is tied to that stimulus. Those percepts must be singular—that is, distinct entities—and whole, not made from smaller parts. You can't mix two things together to get sweet, he points out. Sweetness is a thing unto itself.

Language, he says, supports the idea that there is a limited range of basic tastes. "I think there is lots of psychophysical evidence and linguistic evidence to support the view that sweet, salt, sour, and bitter are generally perceived as whole things, things that are not made up of other things—and that they are perceived as different from each other," says Beauchamp. "Looking anthropologically at words people use for tastes, sweet is, I think, universal. Bitter or bad is universal. Salt is pretty common." The reason so many cultures have a word for these concepts is that they are essential. Everybody perceives them, so every culture built a word for them.

But, Beauchamp says, the body needs more than sugar and salt. We need fatty acids and minerals like calcium, so it makes sense that we might have receptors to help us detect them. Perhaps, he says, it's a detection that takes place physiologically, below the level of conscious thought. Just because there's a receptor for something on the tongue doesn't mean that we have a unique percept for it in the mind. "So Mike will argue that calcium is important, and of course it is," says Beauchamp. "But I would argue that there is no percept that makes it, in psychological terms, a basic taste." In his view, the same goes for fat. While there's evidence that we perceive fat as a mouthfeel, it's unclear to Beauchamp whether fat has any bearing on taste—and if so, whether it rises to the level of more readily distinguished categories like sweet and bitter. So because there is no conscious, unique percept for calcium and fat as tastes, he argues, we've never developed language to describe them. If you don't perceive it, your culture doesn't make a word for it.

But Tordoff thinks the explanation may work the other way around; that is, if you don't have a word for it, you can't perceive it. Or, perhaps more precisely, you don't *consciously* perceive it. "You still get the same sensory signals in your brain," Tordoff says. "You just have no way to interpret them."

For the last several years, Tordoff has been arguing that we can learn to perceive the taste of calcium. Tordoff is a British ex-pat with a neatly trimmed beard and a wry sense of humor, who practically bounds down the hallways as he makes his way from one side of the Monell labyrinth to the other. As Mattes does for fat, Tordoff argues that calcium is necessary for survival, in this case for our bones and for chemical communication between cells. So far, two tongue receptors are known to respond to calcium: the T1R3 receptor, which is also involved in sensing sweet and umami, and CaSR, which is short for calcium-sensing receptor.

(It might seem confusing that a receptor can be involved in sensing more than one basic taste. But it turns out that the T1R3 receptor is only half of a sweet or umami receptor. Through a process called dimerization, T1R3 can combine with other receptors, allowing them to form configurations that fit with different molecules, sort of like locks that can merge to fit different keys. T1R3 combines with the T1R2 receptor to make a sweet receptor, and the T1R1 receptor for umami. Tordoff argues calcium can be sensed by T1R3 combining with another receptor—perhaps CaSR—to form yet another configuration of this lock.)

Tordoff's lab has shown that rodents can discriminate between vegetables with higher and lower calcium contents; they eat the higher-calcium ones when they are calcium deprived, and switch to the others when they are sated. That's the principle of homeostasis, sometimes called "the wisdom of the body." "Basically, when the body needs something it goes out and finds it," says Tordoff.

But when it comes to humans, Tordoff has run into the language problem again and again; he's tormented by the time he got quoted in a news article describing the taste as "calcium-y," a term he came up with in desperation. Still, he has not yet landed on anything better. "Milky" or "chalky" only gets you part of the way there. Maybe you can compare the taste to mineral water, sort of. Now, he sighs, calcium-y is fine. "I don't even try anymore," he says drily.

So unlike the Denver museum's effort to crowdsource descriptive words, Tordoff's lab focuses on discrimination. If people can learn to recognize calcium and differentiate between concentrations, that challenges the idea that calcium triggers an unconscious physiological reaction, but its recognition never rises to the level of perception. Recognition suggests some degree of awareness. "But maybe that is just the very tip of the iceberg of the whole taste phenomenon," says Tordoff equitably. "So we don't know."

Tordoff found that people who are asked to chew through 24 kinds of vegetables can reliably rank them by their calcium content (although when asked to describe the taste, they tend to call it bitterness). And he adds, "It's not easy, but you can train people to recognize calcium taste." In one of his studies, participants were asked to rate the intensity of different solutions. Then, some were given solutions dosed with lactisole, a chemical that hobbles the function of the T1R3 receptor, essentially locking it open and making it unable to bind with molecules. Those who drank the lactisole reported that intensity of the calcium taste decreased; presumably, the locked-open receptor was interfering with their ability to taste it. "That was the fundamental bit of evidence that we have that the T1R3 receptor is involved in calcium taste," says Tordoff.

His lab doesn't have any people training to taste calcium when I visit, but Tordoff figures he can teach me anyway. He escorts me to a lab where his assistant has sorted plastic cups into the compartments of a muffin tin, each cup filled with a clear fluid. "Let's taste some calcium!" Tordoff says gleefully.

The first cup holds a 30-millimolar solution of calcium chloride. (They use soluble calcium salts to make them easy to drink.) "That's about the concentration of calcium that you would find in skim milk," Tordoff says, adding that we can't actually taste the calcium in milk, because it binds to the proteins. "So just drink as much as you want."

I'm expecting, perhaps, Perrier. But it's far more subtle. "It definitely tastes like something, but how to describe it?"

"*Right!*" says Tordoff emphatically.

"There is something about it that reminds me of milk," I say, then sheepishly realize where that's coming from. "Maybe because you just said 'milk' a minute ago."

"People are incredibly susceptible to suggestion," Tordoff replies. "If I was to say, 'Oh, it tastes beefy,' I am sure you would say it was beefy."

We move on to 100-millimolar solution, an uber-dose, about 100 times the concentration found in mineral water. "There is nothing that we would eat or drink that would have such high levels of calcium in it," Tordoff says. I take a sip. It's like licking a battery terminal. "Everything in my mouth just wants to curl up," I manage to report. "My tongue wants to fold itself in half. But it doesn't taste to me like particularly anything. It's hard to think of an analogy."

We try again with a new cup, this time a moderate solution of calcium lactate, another calcium salt. Tastes like tap water, I say with a shrug. Dirty tap water, Tordoff suggests. Maybe a bit of a chalky aftertaste, I say, and then immediately suspect myself for tasting chalk, which Tordoff had also already tossed out as a descriptor.

The last drink is a super dose of calcium lactate. And there it is, just as it was in the fat test: a wave of revulsion, an acid feeling in my stomach. I can't tell you what this stuff tastes like, but I can tell you there's too much of it. Like Mattes, Tordoff thinks there's a reason why too much calcium should be horrible. We need calcium, but too much is lethal to cells. Tordoff thinks we crave calcium when our overall level drops, but instantly despise it if we overshoot.

Tordoff points out that the solutions I have tried are really strong. His lab participants discriminate between much more subtle gradations. But our taste test has neatly illustrated his central problem: On the one hand, it's easy to sense the difference between more and less calcium. On the other, it's very hard to describe, in plain language, what it tastes like. It's somewhere between bitter and sour, but how can I tell if calcium is a separate entity or part of the same band? There is just not a word for it in my rainbow.

THERE HAVE BEEN AT LEAST A HALF-DOZEN CONTENDERS for new tastes since umami. Several labs have tried to figure out whether water might be one. Scientists at the University of California, San Diego have argued that animals can taste carbon dioxide, and that this taste ability is separate from their tactile response to carbonation bubbles. Anthony Sclafani, a psychologist at the Brooklyn College of the City University of New York, has published research showing that rodents can distinguish between three carbohydrate tastes: sweet (for ordinary sugars), large starch molecules, and smaller maltodextrin molecules, produced when those starches are broken down.

And there is, in this search for a sixth, seventh, or eighth primary, a Pluto—a mysterious object that has inspired debate over whether it fits the planetary schema. It's called kokumi, and like umami, the name hails from Japan, although the concept has also been associated with other Asian cuisines. Much of the leading research has been done by Ajinomoto, the Japanese food company already known for its umami research and manufacture of related products, including the flavor enhancer MSG (monosodium glutamate). Translated somewhat vaguely as "yumminess" or "mouthfulness," kokumi is most often described as a quality that enhances other basic tastes, yet has no taste of its own. Many of its proponents—including Ajinomoto—stop short of calling it a sixth taste, referring to it more loosely as an "effect," "phenomenon," or "sensation."

"Kokumi has been used for a long time in Japan as a word to express richness, mouthfulness, thickness, continuity, complexity of the foods. Kokumi was a special culinary word," rather than one widely used among the public, writes Ajinomoto spokesman Dr. Shintaro Yoshida, a veterinary physiologist who speaks publicly about the company's scientific research. The company adopted the word in 1990, he writes, when researchers studying the taste-active components in garlic used it to describe substances that bring out these properties in foods and enhance umami, sweet, and salty tastes. Rather than a sixth taste, he calls kokumi a "flavor modifier," which has "no taste by itself."

The most common description of kokumi is that it's a harmonious characteristic that emerges over time, likely linked with the protein breakdown and marrying of flavors associated with stewing, aging, or fermenting foods. People who study kokumi are looking for it in foods that take time to make: in Gouda, an aged cheese, as well as in meaty broths and fermented products like soy sauce and fish sauce. Several Asian food processing companies, including Ajinomoto, make kokumi-enhancing products, designed to mimic the effects of simmering for foods that are produced quickly.

Ajinomoto researchers believe kokumi is linked to the calcium-sensing receptor, CaSR. This receptor responds to glutathione and several other peptides, which researchers believe are responsible for "kokumi flavor." In tests with live cells taken from mouse taste buds, the cells responded when those peptides were applied, and their responses were blocked when researchers applied a CaSR inhibitor. But rather than sensing a separate taste, the researchers wrote, the actions of these receptors may simply modify the activity of neighboring taste cells or sensory nerve fibers, primarily enhancing sweet and umami tastes.

Ajinomoto researchers have also asked people to taste foods laced with these kokumi-related substances, including samples of umami, sugar, or salt water (representing the basic tastes) or chicken broth. People reported the basic tastes were enhanced when the kokumi substances were present in the flavored water, and that "thickness, continuity, and mouthfulness" increased when it was in the broth. Company researchers have also isolated kokumi-related peptides in commercial soy and fish sauces—significantly, both aged products. A long cooking or fermentation process, Yoshida writes, "seems to increase kokumi substances in foods. However, we are not sure that these processes are the only way to create the kokumi substances."

If you're confused by the prospect of a taste that only enhances other tastes, you're not alone. Many Western food scientists say they just don't get the concept. Yet as strange and wide open as the kokumi idea is, it's worth thinking about, because it forces you to grapple with nearly every big question in taste research: the difference between biochemistry and perception, the idea of conscious tastes and the perhaps subconscious processes that modify them, the complicated web of culture and language that surrounds our efforts to categorize anything. Whatever else kokumi is, it's the perfect example of the language problem in action. The word is familiar to people who work in cuisine or food research, and to almost no one else. It means as little to the world as umami did to Westerners two decades ago. So before I finish my quest to find a potential sixth taste, I hit pause to learn how massively the food world, and therefore the rest of us, had to change before we accepted a fifth.

By pausing, I mean I have coffee with Dr. Ali Bouzari and Kyle Connaughton. The three of us perch around a table at Four Barrel Coffee in San Francisco's Mission District as an endless parade of people in yoga gear and startling facial hair line up at the counter. If you want to see the cutting edge of culinary adaptation, and how popular savory has become, this is the right neighborhood. Long lines to get into an unmarked Chinese restaurant serving ultra-umami Kung Pao pastrami? Sure. Avant-California twists on Hungarian charcuterie and pickles? Yep. Meat-flavored ice cream? Old news. Here at Four Barrel you can also buy very hardy examples of a food trend that shows no sign of cresting: small-batch roasted coffees, the top practically an oil slick of flavor.

Bouzari is a culinary scientist, and Connaughton is a chef, American born, trained in Japan, most recently of England's Michelin-starred The Fat Duck. He's a culinary translator, introducing Western chefs to Eastern techniques. So is Bouzari, who teaches chefs about food science, consulting for high-profile names like the Thomas Keller Restaurant Group. They've been working for the Umami Information Center, a Tokyo-based advocacy group sponsored by the food industry, and one of their projects traces the evolution of umami preference in the American palate. Within this, perhaps, is the story arc of basic taste perception: identification, assimilation, and ultimately, appreciation.

"We talk about umami being new, but Dr. Ikeda's work is from 1908. We just celebrated 100 years of umami," says Connaughton, as we crowd around our little table.

"One hundred years of Japanese people being right," adds Bouzari with a grin.

"Yeah!" agrees the chef enthusiastically. "The science was there. It's undeniable. But it took so long. Why did it take so long to get into the lexicon?"

The story, they think, begins with familiarity, and in Japan that means a broth called dashi. Dashi is "the foundation of all Japanese cuisine,"

Connaughton says. "Miso soup, any sauce, any cooking liquid, any simmering liquid—essentially 90 percent of Japanese food has dashi as the base." It's made by steeping kombu, a sea kelp that is dried and aged to concentrate its flavor, thereby extracting its glutamate. To this you add bonito flakes—dried tuna loin that has been fermented, smoked, and sun aged, producing high levels of inosinate, a meat-based nucleotide detected by umami receptors. "These two things together create this synergistic effect, where the impact of the umami is greater than the sum of its parts," says Connaughton. Basically, dashi is umami turned up to 11. (In his 1909 paper introducing umami, Ikeda even calls out dashi as its exemplar, writing, "The taste is most characteristic of broth prepared from dried bonito and seaweed.")

So if you grew up with a dashi-based diet, you've experienced umami your whole life, and everyone around you has a word to describe it. Cooking around the world, Connaughton has found that cultural tuning can make some diners very sensitive to umami, while others barely taste it: "If I make someone a clear broth of dashi, a beautiful consommé-type soup, the Japanese will say 'umai'—that's like 'delicious.' Umami comes from that word, 'umai.' And Americans will be like, 'Uh, yeah, this is tepid. What is this?'"

That's not to say that people in Western nations don't eat and enjoy umami-rich foods. "We *love* steak," Bouzari says.

"We *love* Doritos," throws in the chef.

"*We are the country that came up with A-1!*" emphasizes Bouzari. "It's not like no American had ever eaten anything that had umami in it. It's just that it wasn't the foundational hallmark of our food culture—"

"—like it is in Japan. Or even Thailand and Southeast Asia," says the chef, finishing Bouzari's sentence for him.

So why did it take steak sauce–loving Americans an extra 100 years to grok umami? Connaughton points out that dashi is simple and concentrated—it highlights umami with a laser-like focus. But in the United States, he says, "the umami that was being eaten in this country was packaged in fat and salt and sweet." Think ketchup and tomato sauce: Even though tomatoes are a glutamate-rich food, commercial versions of these products are rarely aged or slow simmered, which would bring out their umami characteristics, and are often cut with sugar and salt. So even if we love Doritos—("That's like the greatest miracle of food science ever, totally cheating to make the most umami thing possible!" exclaims Bouzari)—it's hard for the untrained tongue to pick out the umami components of nacho cheese flavor over, say, the salty ones.

Yoshida offers a similar take on why umami faced so much disbelief in Western countries. In addition to the decades it took for researchers to produce enough peer-reviewed studies to convince other scientists, "umami taste is also the foundation of most traditional Japanese dishes," he writes. "The clear soup stocks used in Japan with very low fat and mild taste are practically pure umami. And this makes umami taste very noticeable, whereas the Western cuisine [relies] heavily on animal fat and other condiments that mask easily the subtle taste of umami. In this case, umami taste becomes hidden in the background and it is more difficult to detect, without any guidance to make it conscious."

There are other factors you could blame, too, like mistrust of research by a food company with something to gain from selling the world a new product. (Ajinomoto has been selling MSG as a seasoning since 1909, just as it is now selling kokumi enhancers.) There was also the great 1980s scare over MSG sensitivity, now largely debunked. "That definitely didn't help, that the most tangible example of umami was thought to be toxic poison," Bouzari says drily.

And while the discovery of umami receptors was the tipping point in the psychology-to-biochemistry transformation the Monell scientists noted, Bouzari and Connaughton see a second change happening in parallel: worldwide assimilation of Asian cooking. Only a few decades ago in Western countries, sushi was hard to come by. Now it's sold in grocery stores. People who once only went out for Chinese food are trying Korean, Thai, and Vietnamese. "We've had this progression of inauthentic Americanized gateway-drug versions of a lot of these ethnic cuisines," Connaughton says, of people getting used to mall versions of Pad Thai and miso soup. American-born chefs played off that familiarity to experiment with more authentic Asian flavors, to highlight umami without all the salt and sugar of General Tso's Chicken. "We can think something's amazing, as chefs," says Connaughton. But customers get the final vote on what they want to taste. "If they don't get it and they don't like it and they don't enjoy it, it doesn't matter what we think and what we like. It doesn't matter how scientifically on-point it may be—"

"The culture has to be right," Bouzari finishes.

They think a similar acculturation process will happen with whatever the sixth taste turns out to be. Someone will discover a receptor for a molecule, we'll figure out which foods are linked to it, and we'll modify our cooking and eating habits, which in turn will modify our taste perception. Somewhere in there lies the language hack: Once we know the taste, we'll agree on its name, or inventing the name will help us isolate the taste. "I think the chemistry of it will be very clear, and that will come out first," says Bouzari. "What will take longer is the conceptualization of what the experience is."

The two are agnostic on the kokumi question, but agree that there are likely more than five basics. "Looking for just a sixth taste is actually very, very moderate. There are people that are saying to hell with basic tastes," says Bouzari. Like Tordoff, he brings up bitter receptors: Do separate receptors mean we can learn to discriminate the bitterness of caffeine from, say, quinine or alcohol or potassium chloride?

"I think it's like the table of elements—as you discover them, you add," Connaughton says of basic tastes. "You never say you're done. We're always going to discover that next thing." After each discovery, there will be the slow wave of cultural adaptation, they say, led by chefs who literally act as taste-makers, teaching people to perceive subtle differences and appreciate what once seemed bizarre or unappealing. "People in the United States are eating arugula and kale and quinoa because chefs wanted them to," Bouzari says. "Kale used to be *punishment.*"

"Now anywhere you go there is a kale salad," Connaughton adds. The two begin to riff on the foods that were once considered unpleasantly bitter and are now fervently embraced by the hip: Negronis and cocktail bitters. Brussels sprouts. Espresso. "This is a temple to bitter that we conveniently are in right now," Connaughton says, looking at the counter, where a line five deep is waiting for a guy who looks like the Gorton's fisherman to pull them a shot. We start to wonder if the slow roll of cultural orientation toward finer discrimination of bitter has in fact already begun. And kokumi, if it starts to gain traction, may just catch on here, in a place where early adoption is prized. "Surrounded by the future kokumi tasters of the world," Bouzari says, amiably looking around at the crowd.

"With their yoga mats and their Tom's shoes," I say, putting words in a notebook.

"And their bitter enjoyment," Connaughton adds, polishing off his second cappuccino.

"And their proclivity for waiting in line for umami," says Bouzari.

DR. DANIELLE REED'S LAB AT MONELL is one of the first to ask whether Americans are among the future kokumi tasters of the world. Reed is best known for her work on the genetics of sweet and bitter perception; every year her lab sets up a booth at the twin festival in Twinsburg, Ohio, where sibling pairs stop by to donate their saliva to science. In 2013, in collaboration with Dr. Mee-Ra Rhyu, a scientist at the Korea Food Research Institute, they tested to see if the twins could taste kokumi.

As a Korean researcher, Rhyu has been interested in kokumi for more than a decade, and the idea intuitively makes sense to her. When you eat curry or your grandmother's long-simmered stew, she writes, "you easily feel a rich taste that includes deepness, complexity, mouthfulness, persistency, or thickness. It is kokumi, I think." Rhyu believes kokumi likely enhances basic tastes, but is not one itself. (In fact, her work focuses on whether kokumi compounds can be used as salt taste enhancers to reduce sodium consumption.) But she, too, has run up against the language problem. "Kokumi" is a Japanese word and most of the research has been done using Japanese foods, so the term is unfamiliar in Korea, although the countries are neighbors.

Yet Rhyu believes Korea has its own kokumi concept; the analogous word is *gippeun-mat* or "deep taste," which people associate with the aged soybean paste doenjang, some types of kimchi, Korean soy sauces, and most notably Miyeok-guk, a seaweed soup traditionally eaten on birthdays and during pregnancy and nursing because of its high calcium content. "This may be the same in every country," Rhyu writes—people developed their own words for this idea and linked to their nation's typical foods, but don't know how to communicate it to others. Perhaps the concept is hiding in plain sight, obscured by our lack of a common term. "I believe if we find out a universal word relevant to kokumi, we can expect more scientific progress on kokumi research," she writes.

But her American colleague isn't sure what to think about kokumi. "The exemplars of kokumi that I have tasted, I don't get it," says Reed. "To me, it seems like the Emperor's New Clothes. But my Asian colleagues believe in this very much, so rather than dismissing it out of hand, we thought, 'Well, OK, let's get down to it. Let's test it and see what people tell us.'"

Their test was simple, just to see if people could tell the difference between plain popcorn and some with a powdered kokumi seasoning, then describe the experience. And 500 twins later, says Reed, "what we learned from that is that people can definitely taste kokumi—like, people can tell the difference between plain and kokumi. Americans are not super crazy about it on the average, but there is a split. Some people really like it and some people really don't." When the twins were asked to describe the taste, Reed says, "people struggle. They reach for 'savory.' They reach for 'salty.' They also report that it's a flavor, but we don't know what flavor. Some people reach for 'barbecue' and 'cheese.' But it's not like putting sugar in your mouth and saying that it's sweet."

Her lab offered a taste of the powder they used. To me, it tastes *brown*: like pan drippings, the darkest rye, licking the inside of the instant ramen flavor packet. Reed struggles to describe it, too. "When I have kokumi in my mouth, I do get that *hmm* feeling," Reed says. "But I don't have a word for it. And it may be a semantic limitation." Is it hard to perceive something if you don't have a word for it? Reed pauses thoughtfully. "I guess I would ask you why you don't have a word," she replies.

Now, this kind of experiment has a complication. A seasoning mixture is not a single substance, like salt or sugar. Yet with kokumi, Reed says, "trying to isolate the exact components has been a tad difficult." So they used a food product atop another cooked food. Neither of these is "kokumi" in its own right. You taste the enhancer, not the *enhancement.*

If I want to understand kokumi, I need to find its dashi—a real-world food that showcases its essential qualities. But how? Very few Western chefs have heard of kokumi, much less designed dishes around it. The calls I make to Japanese grocery stores end in confusion. And this is when, thank goodness, Fany Setiyo takes me to dinner.

Setiyo is the owner of Le Sanctuaire, which sells spice mixes and culinary equipment to molecular gastronomists and chefs. Born in Indonesia, and a regular world traveler, Setiyo is multilingual in language and cuisine. While she'd never heard of kokumi, the concept immediately reminds her of koji salt, or *shio koji*. This is a Japanese seasoning you put on rice, and although it doesn't have much flavor on its own, somehow the rice just tastes . . . *better.* "It's hard to explain it," she says. "When you taste it, it doesn't taste like anything. But then when you add it to your food, whether it's cooked or about to be cooked, it changes the flavor profile. It's like a magic salt!"

Koji, or *aspergillus*, is a mold, used throughout Asia for fermenting miso, alcohols, vinegars, and bonito. It is valued for its mellowing quality that deepens as flavors marry, which sounds an awful lot like kokumi. There is another clue that Setiyo has made a useful link: One of Ajinomoto's kokumi enhancers is called Koji Aji, and spokesman Yoshida says this is made with—and named after—the koji mold. The mold is often used to ferment soy sauce, thanks to its enzymes that break amino acids into smaller peptides, he writes, and Ajinomoto's researchers have linked soy sauce to kokumi substances. "So, I think koji will be able to increase the kokumi substance in foods," he concludes. Similarly, Rhyu points out that joseon ganjang, a Korean soy sauce, is also made using koji, and this sauce is used in flavoring the seaweed soup she considers Korea's kokumi exemplar.

So if you want to understand kokumi, you might try koji. And if you want to see what koji does to food, Setiyo says, you have to go to Yuzuki Japanese Eatery.

YUZUKI IS A JAPANESE GRILL IN—perhaps not coincidentally—the Mission District. Owner Yuko Hayashi opened the spot in 2011, the first in the United

States to make its own koji, which is handcrafted in a tiny upstairs kitchen. On a hot autumn afternoon, Takashi Saito is readying his latest batch. Saito, who was the restaurant's executive chef until 2014, is wearing cotton chef's whites and carrying a long, white tasting spoon. He speaks thoughtfully, punctuating his words with an emphatic "Hm!" As a sous-chef chops daikon root into minuscule cubes, Saito goes to check in on what he calls his "koji room," a Styrofoam box on a wire shelving unit. (In Japan, koji is grown in much larger actual rooms.) A meat thermometer is pushed through the middle of the box. Inside is a wooden wine crate placed over a bit of water, and nestled on some cheesecloth is a bed of rice, the granules plump and glistening, white fuzz just developing in a few spots.

Saito steamed the rice last night, then inoculated it with the mold, a pale, fine, unscented dust. The temperature inside the box is going up, a sign that the fermentation has begun. He'll control the temperature carefully over the next three days to two weeks, using a silver insulation blanket and ports cut into the Styrofoam. The fermentation mostly makes the rice taste sweet, he says. He uses his long-handled spoon to put a bit in our hands. The grains have a satisfying chew, but still taste like ordinary steamed rice. "Not yet," he murmurs thoughtfully. But it's still early. "When it's done it looks like a blossom, like a mold," he says. "A white beautiful flower."

When the rice is ready, Saito makes shio koji by liquefying it and adding salt. He uses it to marinate meats. The puree is an opaque white; the taste lies somewhere between the tang of miso dressing and the saltiness of lox. He also makes a sweet version, just rice and water. This one's lumpier and it has a soft sweetness, more porridge than candy.

Saito has never heard of kokumi. But his eyes light up as he realizes he knows its root term, *koku*. (Like "umami," "kokumi" is an invented word; the suffix "mi" means "taste.") "We use the word 'koku' at home," Saito says, and then works to come up with a translation. "Koku is kind of a hiding punch," he says, a "back taste" rather than a "front taste." People sometimes use it to mean depth or layering. He brainstorms for a while and then comes up with another metaphor. Think of pillars, he said, which help the taste stand. "For example, like a house. With no pillars, a house is easy to break down. But if there are strong pillars, the house is very strong. Koku is kind of those pillars," he says.

Does he mean that the flavor is unified, it's supported, it's holding together? "Hm!" he says. "I think so." And time, he says, is essential. If you cook a stew for only a half hour, that's not enough time for koku to develop, because the flavors are still too distinct.

So on a Friday night, Fany Setiyo and I sit down to taste this transformation, asking the restaurant's owner to send out whatever best captures the essence of koji. Then we roll tape and try to describe what we taste. The grilled squid, served in thick tranches with a slice of yuzu for squeezing, has a charbroiled flavor from the grilling, and is remarkably soft for seafood cooked over fire. But its most remarkable quality, Setiyo says, is despite the salt in the marinade, you don't taste the salt; "you taste more of the meat." And this becomes our theme. The chicken skewer, with its Thanksgiving turkey–quality juiciness . . . doesn't it just taste *really* like chicken? The salmon, a perfectly pink fillet wrapped in a delicate sheet of cedar and adorned with orange autumn leaves, is intensely . . . *salmon*. The rice, flecked with shaved vegetables— doesn't it taste like brown or red rice, with the nuttiness that usually comes from the bran, even though it is actually white? Maybe, we speculate, koji makes food the platonic ideal of itself. Maybe it just makes meat more, well, *meat-y*.

When I mention this theory to restaurant owner Hayashi, she just laughs. "Some people ask, '*Which chicken do you use?*'" She delivers this in a conspiratorial whisper, mimicking the foodie desperate for the inside scoop on which farm or heritage breed makes for the most succulent bird. "It's not that question that is important," Hayashi continues. "People think the chicken here is amazing. They don't think about koji." So perhaps we have uncovered a clue in the mystery: You have to start breaking food down, like through a koji marinade, to get at kokumi and its related peptides.

I run this past Bouzari, the food scientist. He thinks the reason people trying to understand kokumi keep running across koji is because koji is just really good at bringing out tastes: "The magic of koji is it's a bunch of little happy molecular knives dicing up everything into small bite-sized pieces, so that there can be a greater chance that it will be the right size and shape to fit into one of your receptors." Koji produces enzymes that break down not just carbohydrates, but proteins and fats, he says, "and if you break down any of those three things you are going to get all kinds of tastes and aromas. That's probably the connection."

But there's one more thing. When I'd asked why aging seems so important to kokumi, he'd said that aging brings out the taste of *everything*, because it breaks bigger molecules into smaller compounds. That works for sugars, for glutamate, and it should work for kokumi compounds. "Kokumi may be a product of time," he'd said, "but time does not only kokumi make." So if you want to find a really concentrated, distinct exemplar of kokumi, you should look for it in aged foods. But most of the food at this restaurant is only marinated for a few hours. Whatever molecular change is behind kokumi should be more evident in a longer process. Just like getting to Pluto, the journey to kokumi takes time. There's one last stop to go.

ON A WARM AFTERNOON at The Cultured Pickle Shop in the small East Bay town of Berkeley, California, the steel-and-concrete workspace is awash in the clatter of glass jars and the hum of refrigeration. Here, wife-and-husband team Alex Hozven and Kevin Farley turn out a rainbow of sauerkrauts, kombuchas, and brined vegetables. Hozven, wearing an apron, is at a wooden table in the middle of the room—basically a station wagon–sized chopping block—weighing out curly Armenian cucumbers. On its other side, Farley has taken some kind of pickled root—something yellow and tentacle-y—out of a bed of rice bran, has sliced off a bit with a paring knife, and is staring at it intently.

The couple has never heard of kokumi, although they know koji. Fifteen years ago, before the pickled vegetables took off, they'd experimented in miso-making, inoculating rice, barley, wheat, or soybeans with koji spores, then steaming the grains and controlling the temperature as the mold grew. But unlike shio koji, miso has to sit for a long time—a month to three years, if not longer. That makes possible the transformation of rice and soybeans—not particularly flavorful on their own, Hozven points out—into something new. "The koji allows an unlocking of all sorts of amino acids and enzymes" that change the flavor of the grain, she says.

While the couple gave up on miso as a business model, they still use it and other koji-related fermenting practices in their shop. They've even kept some of their original ferments around. And because of this, Farley has put aside the tentacle and started shuttling in and out of the walk-in refrigerator, emerging with glass jars filled with brown pastes, scooping their contents onto a dish in four little clusters.

He's figured out a way for us to taste time.

We're going to taste the koji ferments from youngest to oldest. He indicates a dab the color of oatmeal, the youngest on the plate, only a few weeks old. "This is kasu, which is sake lees," he says, or the sediment left over after the rice used for sake-making has been inoculated with koji and fermented. These lees come from a nearby sake factory. To this putty, Farley adds sugar, salt, and the vegetables he wants to pickle, and the fermentation process continues for a year or more.

The second bit of kasu is the color and texture of mashed banana. This one has been fermenting for six months. Next to it is a barley miso—at two years old, it's sienna brown. Farley indicates the darkest, thickest paste. "This here is one of the last relics of our miso," he says. Now 12 years old, it's almost black, like deeply caramelized onions.

Farley hands me a fork, and we start with the youngest, palest ferment. It is sharply sweet and salty, first one, then the other. The next one, with nearly six more months to mellow, has a fruitier quality, and the taste, rather than a biting attack on the tongue, has become a whole-mouth experience that fills the top of the oral cavity and the back of the throat. "Most of that sugar has been metabolized out and the salinity becomes very soft and round," says Farley.

We move on to the two-year-old miso, and the flavor is deeper still. And when we try the 12-year-old miso, my first surprised reaction is that it tastes like steak. The flavor is dark and thick. Perhaps I'm borrowing the association from Worcestershire or soy sauce, both aged and used for flavoring beef. But it's clear that we're witnessing the unfolding of a long chemical process, and that what koji does, it does over time. "The flavors are less married in the younger and are married completely in the older ones," Farley says. In the younger ones, the sweetness and the saltiness are separate, rolling through in waves. With the older ones, the flavors have become more integrated and pungent, swelling higher into the nose and deeper into the throat. This, I realize, might be what the Ajinomoto researchers mean by continuity and "mouthfulness."

Hozven leans in with her own fork. "It is really hard to describe taste. It's super hard," she says, echoing the plaint of every taste researcher on the planet. But she gives it a try. Hozven's descriptions are nearly all metaphors for shape and motion. She describes the youngest kasu as "flatter" than the others. With the older ferments, she says, the taste "becomes much more circular. It sort of weaves its way around. It takes you on a trip." The older ones "taste whole," she continues. "It's not just a sweet, sour, salty. It makes you think about it for a long time."

Because the couple is curious about what I'm after, I tell them about the hunt for a sixth basic taste, the language problem and how hard it is to isolate a sensation when you don't have a preexisting concept for it. How do you know what you're tasting, I ask, until you have a word to describe it?

"I guess I don't really feel about it that way so much," Hozven says thoughtfully. She doesn't mind if we can't find the words, if the closest we get is metaphors. She knows what she's tasting because *she is tasting it*, not because a scientist associated it with a receptor or a word or a perceptual category. And for someone who has worked with pickles as long as she has, time is just something you cook with. Maybe that brings out kokumi, or maybe something else. Whatever it is, the practical experience of cooking and the personal experience of eating interest her more than giving a taste a precise name or a number. The rainbow's what matters, not how you dissect its stripes.

"Time is its own ingredient," she says. I press her to tell me more, and she laughs gently. "I mean, that *just is*. I sort of enjoy not even breaking that down. It sort of speaks to the alchemical nature of the process in a way that we can witness and delight in and never fully understand. And that's OK."

Then she has to get back to the cucumbers, so I turn in my fork and pack up my notebook. On the way out of the shop, I call over my shoulder, "I'll let you know if I solve the mystery."

Hozven smiles and shrugs. "I wouldn't really want you to," she says.

TWO

Smell

IN A LITTLE UPSTAIRS ROOM at the French hospital Ambroise Paré, Anne Ca-milli unlocks a refrigerator. From within its cool depths, she withdraws a box. Inside the box is a plastic bag. Inside the bag are more tiny bags. Inside each tiny bag is a square glass bottle, no bigger than a postage stamp. And inside the bottles are memories.

Well, technically, inside each bottle is a fragrance, the essence of a fruit, flower, or beloved French foodstuff. But at the Atelier Olfactif, the smell work-shop she is about to host, these fragrances are just a means to memory's end.

Working quickly, Camilli begins uncapping bottles and giving each a sniff, making sure the fragrance hasn't turned, or that it isn't what she calls "too complicated." To work, the fragrances must be easy to identify and capable of stirring warm recollections. "Rhubarb. Very typical," she says, inhaling deeply from a bottle. "Have a smell. Really, it reminds me of England, rhu-barb pie with custard. I remember this. I close my eyes; I see myself in England when I was 12."

Satisfied, she places it in a carrying tray, then jots its name into her rec-ords, purple ink in a purple notebook. Once there are ten bottles lined up, she adds a package of blotters, the long paper wands perfumers use to waft scents under the nose. She is ready to go.

Camilli once worked for a company specializing in perfume creation and natural raw materials—sandalwood oil, patchouli leaves, vetiver grass—where she learned to "train her nose," to remember each scent, to differentiate be-tween two crops of the same flower or spice. She later worked for a cosmetics packaging design studio, and now runs a communications agency that repre-sents perfumers, product designers, and other creative people. She's a lively woman wearing a chic navy pantsuit, her dark hair cut into a long bob. She's here today as a volunteer for the atelier, a first-of-its-kind project run by a non-profit called Cosmetic Executive Women.

Since 2001, the group has been bringing "olfactotherapy," or scent therapy, to patients in 16 French hospitals, the wards where the people are dealing with the loss of smell as a side effect from head trauma, stroke, or chemotherapy following cancer. But Ambroise Paré, on the border of Paris and its western suburbs, has very special patients with very special smelling problems. This atelier works with the geriatric ward, and many people treated here have Alzheimer's disease or another dementia.

The tiny bottles Camilli is carrying are meant to offer a rare moment of relief from the steady corrosion of memory that Alzheimer's causes as it gnaws its way through the brain's circuitry. It is a progression that starts, curiously enough, at the very top of the nose. Olfactory loss is the first clinically diagnosable symptom of Alzheimer's. It is also an early symptom of other neurodegenerative disorders, such as Parkinson's, although olfaction's link to Alzheimer's is so far the most studied.

The elevator is taking too long, so Camilli hurries down the stairs, explaining as she goes that she's never sure whom she'll see at each workshop—it's a short-stay unit and people arrive after a variety of medical mishaps. But in general, the people are between the ages of 80 and 100, and frequently have a cognitive disorder. Sometimes they come to the hospital following a more commonly recognized symptom of dementia—a fall, trouble remembering where they are—and are diagnosed later.

No matter; she just wants them to smell, remember, and enjoy. This atelier's approach to sensory function couldn't be more different than the grinders': The atelier's motivator is loss, not enhancement, and their attitude is patience, not frustration, as befits an organization working with a currently incurable disease. But they, too, know the art of the hack. While they can't hurry up an antidote for lost memory, they can do the next best thing: bypass damaged connections.

Camilli sails into the room where the patients are gathered, warmly greeting Dr. Sophie Moulias, the geriatric ward's director, and Laure Pellerin, a physiologist who will help run the workshop. Five patients, a man and four women, are seated around tables in a sparsely furnished room. (For privacy, all medical patients in this chapter will be identified by pseudonyms.) They regard the bottles with friendly curiosity.

"Do you know what we are going to do today together?" Camilli asks brightly.

"Oui?" says someone tentatively.

Camilli laughs, and tells them she is here to make them "work" their noses, because odors are a valuable part of life, of the enjoyment of food. Today they will smell fruits—good scents, she promises. As she talks, Camilli dips the blotters into the first bottle, offering each with a polite "Madame," demonstrating how to wave them gently under the nose, provoking a scented breeze. "Does it smell good to you?" Camilli begins. There is a chorus of yeses, so she presses further. "Does this odor remind you of something?"

There is a long pause. "An orange?" ventures Josephine, a sweetly impish woman in a teal nightgown seated in the pushchair to Camilli's right.

"You're not far off," says Camilli encouragingly. "It's in the same family. Very good."

Moulias leans over to Julien, the only man in the room, a trim figure in blue pajamas. "Does this remind you of something? Of a particular fruit?"

"No," he says quietly.

Puzzled silence falls over the room until Eleanor, a kind-looking woman wearing a brightly colored scarf that only partly covers the sling on her arm, hazards a guess. "Lemon?"

"Ah yes, lemon. That's it!" announces Camilli. As she gathers up the used blotters, Camilli chats with each patient, asking them if they wear perfume or cologne. She's making conversation, but she's also trying to warm them up to the idea of smell and personal associations.

The next scent goes around the table. Camilli offers clues: It's a year-round fruit. It's sold at every grocer. It's European, not exotic. But the group is stumped. The doctor leans toward Julien again. Does this remind him of anything? "No," he replies flatly. So Camilli keeps prompting. There are many varieties of this fruit, some sweeter or more acidic than others. Its exterior is red. At this, Eleanor makes the connection: It's apple.

It's readily apparent that there's a wide spectrum of smelling sensitivities in the room, as well as states of cognitive health. Eleanor, the best guesser, is clearly here for her arm, and the doctor thinks she has no cognitive deficits at all. For Josephine and another patient, a woman who will sit wordlessly throughout the entire session, clutching her purse on her lap while smiling amiably across the room, the doctor suspects some degree of impairment. Both came in for other maladies, and haven't been tested yet for cognitive health. But Julien has Alzheimer's, she says, and so does Delphine, a tiny, gaunt woman curled up in a wheelchair. So far, Delphine has been silent, sometimes dozing off, despite the efforts of Pellerin, the physiologist, who is seated close to her and has been gently waving the blotters under her nose. Delphine's case is advanced, says Moulias. Like many people with dementia, Delphine is withdrawn and uncommunicative. Nobody in the ward has heard her say more than "yes" or "no" since she got here.

Camilli begins the next round, this time with the rhubarb. It's a stumper, too, and the room goes quiet as everyone sniffs. Then from Delphine's corner of the room, there is a stir. She has woken up, smelled the wand, and recoiled in dismay. And yet she is wearing an enormous smile, an expression somewhere between annoyed and amused. Camilli and the hospital staff exchange delighted looks. The mood in the room warms. People are getting the hang of the game.

So Camilli keeps going, passing around the next scent, almond. And Delphine, now leaning forward in her chair, staring straight at Camilli, hates this, too. As the other people in the group chat animatedly about the scent—"*un peu amer*," a little bitter, they agree—and about almond cake and Amaretto, Delphine is leaning close to Pellerin, who bends over to listen. The older woman's arm points at Camilli, and the physiologist translates her request: *Next time, give her something good.*

FRANCE, THE COUNTRY OF PERFUMERY and Marcel Proust, knows something about the power of scent to trigger not only memories, but emotional responses. Proust gave us literature's most famous example in the first volume of *In Search of Lost Time*, when he described how dipping a madeleine biscuit into a cup of linden, or lime flower, tea flooded him with the memory of a childhood trip to his aunt's house in Combray. Scientists have dubbed these emotionally potent recollections of lost time "Proustian" in his honor.

Smell is a primitive sense, an alert system that lets creatures know when they have stumbled across dangerous or desirable chemicals. Because olfaction developed so early in the brain's history, it is intimately connected with the centers for memory, learning, and emotion, which piggybacked atop it. It is a way we link the sensations of our present to the experiences of our past. And when those links are damaged, we forget.

Alzheimer's is the antimadeleine, the unraveling of ties between memories and the scents that bring them back. But the disease's attack is subtle; it doesn't completely wipe out your ability to smell. Instead, it slowly erodes your ability to distinguish one scent from another. It makes madeleine and lime flower and beef stew and cigarette smoke and rotting garbage all smell, eventually, more or less alike. It prevents you from putting together tea and biscuit and getting Combray. Without that distinctive combination to recall the summer with your aunt, your connection to that moment fades.

But don't panic: Some smell loss is normal. Olfaction is sharpest among the young, and dulls thanks to a lifetime of wear and tear: colds, allergies, breathing in pollutants. The sensory cells in the nose are remarkably resilient against damage; their population turns over every 28 days. (Taste receptors are the only other sensory cells that regenerate.) But even they have their limits.

Alzheimer's does more than accelerate or echo the slope of normal olfactory decline; it is a cascade of pathologies that, after appearing in the olfactory tract, proliferate to damage increasingly deep systems in the brain. There are some 47.5 million dementia patients on the planet, the majority of whom have Alzheimer's. With diagnoses expected to nearly triple by the year 2050, and no cure in sight, medical experts are hoping that the world of scent will offer clues about how to manage this disease. Some think that scent could be a powerful tool for early diagnosis, and that intervening earlier might slow dementia. Others wonder if, for already severely affected people, scent can offer memory therapy. Even after Alzheimer's muddies the ability to distinguish between odors or recall their names, scent-based memories still linger in the mind. If the direct approach to the past no longer works, can it be summoned another way?

When Marie-France Archambault, a now-retired perfume company owner from the south of France, came up with the idea that would become the Atelier Olfactif, her mission was much broader than Alzheimer's. Her Cosmetic Executive Women group initially just hoped to inspire a moment of pleasure for long-term hospital patients tired of bad cooking and medical smells. They partnered with fragrance industry giant International Flavors & Fragrances, which creates the scents used in the workshop, and brainstormed what might appeal to cloistered patients: the sweet smells of childhood, fresh outdoor scents, appetizing foods. Some of their choices are incredibly clever and subtle, although perhaps obvious to cosmetics executives, like the scent of lipstick. Nearly everyone has a positive childhood memory of lipstick, Archambault says, because they associate it with their mother's kiss.

But Archambault, who often works with the program's geriatric and trauma patients, soon realized how evocative scents can be, and those arms of the program became reoriented around memory. Even with Alzheimer's, she says, people can experience very intense recollections. "Odor is always related to a place, a person, and something you were doing at that time," she says. "I was in the garden, I was at the sea, I was in a boat, I was with my grandmother, I was with my best friend." And most important, she says, "It is always related to emotion. Always, *always.*"

Sometimes those memories are general, reminders of holidays or youthful habits. Among her clients, says Camilli, the scent of walnuts tends to bring up fond memories of the "king cake" served for the Catholic holiday of Epiphany, or of a white glue with a nutty smell that French kids remember from school. Women often link floral scents to perfumes they wore on special occasions. Mushroom and berry smells remind people of rambling in the woods.

But sometimes people recall a vignette. "One day I had this very tiny old lady, all dressed in black, very witty, 99, about to be 100," says Camilli. "I had them smell flowers that day and they were on orange blossom. And she closed her eyes and she said, 'This reminds me of Tunisia.'" The woman had lived there as a teenager, and told Camilli about going to a village where they harvested orange blossoms, recalling how boys and girls were not allowed to walk on the same side of the street.

Moulias fondly remembers a man with advanced Alzheimer's who, like Julien, struggled to smell anything. "Each fragrance, he said nothing. *Nothing.* Not 'I don't smell,' or 'I don't know it.' Not a word," she says. Then they gave him the scent of vetiver, which is often used in cologne. "Ah!" he said, suddenly animated. "How many girls I seduced with vetiver!" Moulias bursts out laughing at the memory. "He said that, you know, very like a young boy."

Archambault recalls a former banker, in the later stages of dementia and long uncommunicative, sniffing a wine-scented blotter and immediately launching into a lengthy story about a tasting tour with his wife. A younger man, who had suffered serious head injuries in a motorbike crash, had been almost wordless until they gave him a blotter laced with the scent of asphalt, upon which he burst out with, "My accident. *Moto.* Dead." (Not all memories, she points out, are good memories.)

With scent memory, she says, "You open a little drawer, and there is a little story." And with dementia patients, after that story spills out, the drawer is shut again very quickly. "Maybe five minutes after you say, 'Do you remember you spoke to me about that?' 'Oh no, I don't remember.'"

She pauses and I can almost hear her shrugging. "That's it."

But these breakthroughs are important, and that's why the group at Ambroise Paré encourages family members to sit in on sessions. Perhaps a moment of lucidity will help them connect with their loved one. "They see that the patient could be different; he is not always the same," Moulias says. "It's a man with his own history, his own ideas, his own taste." Despite the dementia, the person that they love is still in there. Scent memory is a way to get them to come out.

And so as Delphine now watches attentively, Camilli's workshop rolls along. The mood in the room has become quite merry, even for Julien, who says he can't smell a thing but seems to be enjoying the company. The others have sniffed their way through wild strawberry, pear (which everyone got fairly quickly; it's the scent of a familiar French liqueur), black currants (ditto; it's the basis of cassis), coconut (which puzzles everyone), and blackberry, which Josephine nails right away, with evident delight.

"Madame, I suggest you train and apply for a job as a perfumer, because frankly you have a super nose," Camilli teases her gently. "A new career," jokes Josephine, giggling adorably.

They are almost done. Delphine is now leaning forward, chin in hand. "An odor that smells good," Pellerin reminds Camilli on the older lady's behalf.

"That smells good?" Camilli asks, laughing. "I'll try, but it's subjective."

She readies blotters as the hospital staff chats about Delphine's transformation. Camilli leans over to make sure I'm getting the translation. "They are saying this woman, she is really having difficulties in walking and with expressing herself generally, and that they have never seen her like this."

One bottle left. "A summer fruit," says Camilli, urging people to close their eyes. "Peach?" Eleanor guesses. No, bigger, hints Camilli. It can be an entrée or a dessert, adds the doctor. The answer begins to dawn on everyone, the delighted reaction traveling in waves around the room. It is a favorite fruit in France, particularly for people who have spent time in the south, and it's synonymous with warm weather and sweetness. People are happily waving their wands and making contented "mmm" noises.

Pellerin bends close to her patient, offering her the paper strip.

"It smells?" she asks. Delphine's lips just barely move.

"Oui," comes a very soft voice. "*Melon.*"

SMELL AND ALZHEIMER'S RESEARCH INTERSECT at the study of memory, and the work done at this crossover point has illuminated much about normal olfaction, and what may be out of order when it malfunctions. You might be surprised that scientists haven't had this dead to rights for years, but olfaction experts sometimes despair that their field is decades behind the other sensory sciences, perhaps neglected because of negative cultural associations with bad smells, or because it's hard to quantify odors, or because the tongue has stolen all the glory for the sensation of flavor, despite the nose's overwhelming role. That's an ironic fate for what is probably life's oldest sense. In the beginning, smell didn't just ensure we could remember a story. It ensured we lived long enough to have one.

Unlike taste, which requires you to put your environment into your body in order to understand it, smell works at a distance. It clues you in to who is nearby, and whether you should flirt or flee. It entices you toward tasty food and repels you from the poisonous or spoiled. It tells you where you are and what's coming at you. "Being able to detect a smell amounts to having a way to predict upcoming events in the environment," says Dr. Jay Gottfried, a neurologist at Northwestern University whose lab studies odor perception.

Researchers believe that a "chemical sense"—a predecessor to smell and taste—was likely the first to evolve, giving creatures as primitive as bacteria hints about what to approach or avoid. (In the lingo of cognition, that's "go" and "no-go" information.) Our smelling apparatus has gotten more complicated over the millennia, plus we've evolved a memory function that helps us remember which is which.

People carry genes for around 1,000 odor receptors, of which only 350 to 400 are expressed in the average person. These influence how many active receptor types you have, and how many of each kind, making everyone's sense of smell unique. But even the best sniffers can't smell everything. Odor molecules must be small enough to become airborne, they must be able to bind to receptors in the nasal passages, and there must be enough of them for us to pick up the trace. Rodents, dogs, and some insects can sense odors in much lower doses, which is why a bee can sniff a bomb, and you can't.

When you inhale, odor molecules bind with some of the 20 million sensory neurons in your olfactory epithelium, a mucous membrane deep inside the nose. These neurons send information to the olfactory bulbs, one per nostril, located roughly behind the eyes, like a pair of fuzzy dice hanging on the nose's rearview mirror. The bulbs feed information to several regions in the forebrain, most notably the piriform ("pear-shaped") cortex, which does the heavy lifting when it comes to identifying odors.

Most scents are complex chemical compounds; there is no single molecule for the smell of, say, buttered toast. "Chocolate probably contains 500 unique odorous components," says Gottfried. Given the hundreds of molecules that can code for a scent, and the many receptor types available, the combinatorial possibilities are enormous. The brain deals with this math problem in a clever way: Instead of making a one-to-one connection between molecule and receptor, as taste cells do, many neurons work together to code for each odor. Their activation creates a three-dimensional pattern that is unique both spatially (the layout of where those neurons are) and temporally (the order in which they engage.) You can think of these patterns as topographical maps: In one, the points might be arranged like the tall, jagged peaks of a mountain range, and in another, the pattern might be more like rolling hills. "Imagine a topographical map of how Kansas would look—maybe that's mint," says Gottfried. "And then a map of how Idaho looks and that would be lemon."

In fact, the same neuron pool can map both mint *and* lemon. This constellation of neurons will activate, but in a different arrangement, with each neuron responding at a different intensity. This flexibility allows us to recognize a huge range of odors, at least tens of thousands. In 2014, researchers at

The Rockefeller University in New York released a study claiming the number was over 1 *trillion.*

The anatomic position of this olfactory processing is also special. Unlike the other senses, smell is only a few synaptic connections away from the brain's emotion and memory centers, which gives it a special poignancy. Think of a scent that brings back a sweet childhood memory, maybe the baking odor of your favorite cake wafting through the house on your birthday. Olfactory information doesn't make a pit stop at the thalamus, which acts as a first way station in all other sensory processing. Instead, when you inhale that cake scent, the information goes directly from the olfactory bulbs to the limbic system. That's where the amygdala controls emotion (your happy feelings connected to the cake) and the hippocampus aids learning and memory (your recollection of past birthdays). Nearby, the ancient entorhinal cortex (literally: "inner nose") is also playing a role in memory and sensory processing.

"It's like this pure information from olfaction going into this part of the brain that is emotion and associations," says Dr. Rachel Herz, a Brown University psychologist and cognitive neuroscientist who specializes in the study of odor and memory and is the author of *The Scent of Desire.* You smell cake, you get instant warm fuzzies.

As the human brain evolved, vision became our dominant sense. The visual cortex grew, the olfactory center shrank, and many of olfaction's alerting functions were traded to the limbic system. Today, Herz argues, for us, emotion has replaced the role smell plays for animals. "Smells tell us about danger, smells tell us about love, smells tell us about these really basic sort of go, no-go things," she says. "And if you think about what our emotions do, they do the exact same things." Emotions help us learn the value of each scent, because we make a positive or negative association based on what happens next. Do you get sick after drinking from that slow-moving stream? Better avoid brackish smells. Does the vetiver cologne seduce the girl? Now that scent is connected with young love.

Associative learning is a plus for a "generalist" species like ours that has put down stakes all over the planet, because that lets us adapt to new conditions and locations. "Humans and cockroaches and rats, we are capable of exploiting any habitat that the Earth has to offer," says Herz. "As a result of that, we also are faced with this pantry of unending possibilities of things that we can eat—and who might want to eat us." If our first association packs a powerful emotional punch, we won't make dangerous mistakes twice. People also learn from the mistakes of others. If we see others disgustedly grimacing at—or spitting out—something with a foul odor, we should get the clue to avoid it.

In fact, scent memories can be so overwhelming that Herz and her colleagues wondered if smell actually prompts more accurate recall than other sensory cues. In an experiment, they found that subjects rated the memories recalled by either photos or odors of childhood objects like crayons, Play-Doh, and Coppertone sunscreen equally vivid and accurate. But the scent memories, the subjects said, were more emotional and evocative, making people feel like they had been transported to a distant place and time.

Building on this, Herz's group ran a second experiment using three nostalgia-inducing cues: popcorn, campfire smoke, and freshly cut grass. Based only on the name of each cue, participants were asked to think of a memory and rate its vividness, emotionality, and evocativeness. They next sniffed, saw, and listened to a sensory cue for each of the items, recalling their memory and rating it again. For example, they saw a video of a campfire, heard a clip of one crackling, and inhaled from a jar of campfire-scented oil beads. Once again, memories recalled by olfactory cues were judged more evocative and emotional than the others, but no more vivid.

Smell triggers the most sentiment when it activates a seldom-retrieved memory, often via a rare mix of scents, like Proust's madeleine mingling with the linden tea. "When we have flashbulb-kind of Proustian memory experiences, one of the things that make them special is the surprise factor," says Herz. "They've unlocked something they might otherwise never have been remembered." These memories are not the cherished ones you have purposefully relived, nor are they cued by simple, everyday scents. They are the complex aroma of the long-forgotten baby blanket rediscovered in the attic; they are the whiff of your grandfather's aftershave and pipe tobacco on a passing stranger; they are that adulthood moment of returning to your elementary schoolyard and inhaling the funk of playground tarmac and wet autumn leaves and cafeteria Sloppy Joes.

Part of their magic is that because these blends are so singular, your memory of them is free from interference. "The smell of coffee, let's say, is connected to so many experiences it's very unlikely for it to be able to unlock a very specific memory," says Herz. "But a unique combination of food odors and outdoor smells could be the smell that suddenly reminds you of that one time at camp when you were 12."

So that's what's going right when your sense of smell sends you back in time. Now let's return to Alzheimer's, where it starts to go wrong.

OLFACTION'S DEEP CONNECTION TO MEMORY makes it uniquely vulnerable to memory's diseases. Or maybe it's the other way around. Maybe something about olfaction invites and spreads Alzheimer's.

Scientists have long known that smell loss is an early symptom of the disease. In their groundbreaking 1991 paper describing its staging, German anatomists Heiko and Eva Braak noted the first appearance of Alzheimer's pathogens in the entorhinal cortex, as well as the transentorhinal region, which connects it to the hippocampus. (Similarly, in their 2003 paper describing the staging of Parkinson's, the Braaks and their colleagues noted that one of the areas where lesions first form is the anterior olfactory nucleus, which lies between the olfactory bulbs and piriform cortex.) Doctors have observed that while people with Alzheimer's lose cognitive function, in most cases they can see, hear, and touch normally. (Taste can fade as smell dulls, but not because anything is wrong with the tongue.) The olfactory loss can't be attributed to the neural die-off that accompanies normal aging—it's too rapid, too devastating. And it not coincidentally maps the progression of Alzheimer's disease through the brain.

Why that's happening is the zillion-dollar question that Dr. Daniel Wesson, a neuroscientist at Case Western Reserve University in Ohio, and his lab are trying to answer using mice. Rodents make a good lab model because their olfactory systems are like ours, and it's easy to breed mice with certain genes strategically expressed. Because they smell for survival, you don't have to train them to sniff. They'll just do it.

Wesson is a friendly young assistant professor with a sandy buzz cut, and he's poring over notes he kept as a postdoc back in 2008. At that time, people had used mice engineered to express Alzheimer's genes to show that they would develop brain pathologies and have problems with learning and memory. "But no one had actually seen if they recapitulate this very early feature in Alzheimer's, which is the smell loss," Wesson says. So he and his colleagues gave it a try.

Now before he gets too deep into the experiment laid out in his notes, we should have him unpack a few basics about Alzheimer's. This is a field that currently offers more riddles than answers. There is no consensus on what prompts the complex cascade of events that lead to symptoms, or on the exact mechanism by which neurons become impaired. Several genes have been linked with Alzheimer's, but none is a smoking gun. People without any of the genes sometimes develop symptoms, and people with them can have normal health. Environmental triggers—brain injuries, stress, previous illnesses—may play a role in whether those genes are expressed among carriers, but their influence is unpredictable. Even diagnosis isn't a sure thing. The gold standard is still autopsy.

When pressed for a culprit, most researchers will finger two suspects: plaques and tangles.

Wesson focuses mostly on plaques, the buildup in the brain of a peptide called amyloid beta. Plaques seem to start with the amyloid precursor protein, which is an intrinsic part of all neurons, although its function is not fully understood. This protein, a wiry filament like a microscopic noodle, is chopped into fragments by enzymes within the cell, and these bits become soluble amyloid beta, which floats around in the brain. At some point, a neuron can become overloaded with it. Maybe it's overproducing it. Maybe it's not clearing it fast enough. Maybe the amyloid is too fragile.

Whatever the case, when the neuron fires to communicate with its neighbor, it can spit soluble amyloid beta peptides out into the matrix that surrounds the brain cells. These glom onto the outside of neighboring neurons, forming sticky sheets called plaques. Neurons exist to chatter; now it's harder for them to talk—or to listen.

The plaques can also just sit out there in the matrix, becoming speed bumps on the brain's information highway. Your neurons can reroute information flow around them, but at a cost. "The precise timing in the way that neurons fire is really critical," says Wesson. "Now you are altering that precise timing. And if that happens not just in one neuron, but in millions and millions of neurons, that could cause some problems."

Then there is the second suspect, neurofibrillary tangles. These tangles are "a fundamental pathological hallmark of Alzheimer's disease," says Dr. Wes Ashford, a psychiatrist who studies aging and Alzheimer's for a clinic run by Stanford University and the U.S. Department of Veterans Affairs.

Once again, the root of these problematic tangles is a naturally occurring protein: tau. Tau is involved in the structure of the neurons themselves, and helps transport materials throughout the brain's neurons. But if these proteins pick up an extra phosphate group, they start to clump together to form clusters of filaments, then threads. These, too, disrupt a neuron's ability to talk to its neighbors. Neurons communicate by forming connections with each other called synapses. These are activated when the neurons' axons, which are like tiny branches or arms, signal to their neighbor through a kind of electrochemical handshake. An axon's signals are picked up by the next cell's dendrites, which are shaped like little leaves or fingers. The threads, says Ashford, "can amputate axons and dendrites," which quashes their ability to function. These amputations compromise the stability of the affected neurons, which is "just like cutting off the limb of a tree," says Ashford. Lose too many branches, and the tree begins to wither. So with the massive loss of synapses, he says, dementia begins.

The threads, he says, can "also be sucked back to the neuron cell body to form neurofibrillary tangles, which is the visible evidence of the Alzheimer pathology associated with the development of dementia."

Scientists are puzzling over why Alzheimer's pathogens first appear in the entorhinal cortex. Perhaps it is because it's incredibly active in the formation of new memories, constantly forming and destroying synapses, continually adapting to and deciding whether to remember incoming information. As Ashford points out, Alzheimer's "attacks fundamental processing of new information and memory, known as neuroplasticity." So perhaps it's not surprising that the most plastic parts of the brain are the most vulnerable to Alzheimer's.

The olfactory system is also always on—we breathe constantly, so those neurons must chatter nonstop—and a high level of activity could also help problems spread. "The more a neuron is active, the more likely it can release some of the pathogens, especially amyloid beta," says Wesson—and the more likely the next neuron will pick it up and do the same. Wesson notes that the olfactory system is structured like a road, which splits into many others, each capable of transporting amyloid. "From one site, information can go out and be distributed throughout the brain," he says. Other theories are that the region acts as a sink for the buildup of amyloid from other parts of the brain, or that damage to the area caused by injury, diseases, or inhaling toxins can trigger neurodegeneration in people who are genetically primed for it.

No matter how the problem starts or which pathologies are at fault, the damage compounds with age. "Age is the biggest risk factor for Alzheimer's," says Wesson, "because the older you are, the more likely it is that something could have happened and things have accumulated." Current theories suggest Alzheimer's is likely a "multi-hit" process, in which the brain endures insults until it reaches a tipping point at which the progression of damage becomes pathological, not normal aging.

And this brings us back to Wesson's notes. His first experiment was just to see if the presence of plaques correlated with how well mice can habituate to odors. His idea was that a healthy mouse only has to smell an odor once to recognize it; if you keep giving it the same odor, the mouse will sniff less and less until it no longer has to sniff at all. But Wesson used genetically engineered mice that at only a few months old had brains already showing amyloid beta aggregations. "And what you can actually see in the data," he says, turning to the notes, is that even after repeated exposures, "the animals kept sniffing for more than they should." They were having trouble recognizing the scent.

Next, he wanted to see if the problem got worse as the gene-altered mice aged—and it did. The older they got, the longer it took them to habituate, especially compared to their "wild type" (nonengineered) counterparts. This correlated with the spread of amyloid through the brain. On autopsy, mice that were only three months old—the age when they began to show olfactory deficits—already had amyloid deposits in their olfactory bulbs. In older mice, the amount of amyloid had grown and moved into other brain structures, like the piriform cortex. This showed that aging and difficulty with smell correlated with the spread of plaques through the mouse brain. And this, Wesson thinks, is likely what is happening with people, too.

You can't autopsy people in early stages of Alzheimer's, but you can show that they have trouble recognizing and habituating to odors. Gottfried's lab at Northwestern, for example, asked both mild-stage Alzheimer's patients and healthy people of the same age to inhale scents while lying inside an fMRI scanner. This machine images brain activity by tracking blood flow. The subjects were given four odors—two minty, two floral. While both groups rated the odors similarly in intensity and pleasantness—indicating they felt equally sure that they had smelled the scents—the Alzheimer's group had trouble discriminating not only between the two kinds of mint and floral, but even between the two categories. The scans also indicated that in healthy subjects, the piriform cortex was less active when encountering an odor similar to one previously sniffed—it recognized the familiar. But with Alzheimer's subjects, there was less adaptation, just as with Wesson's mice.

Wesson's early studies were behavioral, but he soon began outfitting his mice with electrodes so he could record their brain activity as Alzheimer's progressed. Again, he and his colleagues found that as the mice aged, their brain activity became increasingly abnormal. He pulls up a graph showing the theta oscillations recorded from the olfactory tubercle of an Alzheimer's mouse. These waves are a signature of brain activity—they rise and fall as the mouse inhales, and their size increases each time it smells an odorant. Their curves should be smooth and sinusoidal, but these are jagged. Wesson thinks that at this early stage of disease, the olfactory bulb actually becomes *over*active, trying to transfer too much information to the olfactory cortex, distorting the signal. "It would be like being in a room with 100 people who are trying to talk to you," he says.

At this stage, his electrodes show that activity in the hippocampus is still normal; that's why the mice only have problems with smell, not learning or memory. But they will soon, and Wesson believes that's because the overactive olfactory bulb is rapidly pumping amyloid into "downstream" structures like the piriform cortex, entorhinal cortex, and hippocampus. Around age seven months, this hyperactivity dies down, but by then the pathogen has spread. Now downstream areas will become *less* responsive as plaques develop. And both problems—too much and too little activity—throw off the delicate neural timing needed to recognize scents.

When I visit his lab, neuroscience graduate student Kaitlin Carlson and postdoctoral scholar Marie Gadziola are working on its next generation of projects. Gadziola gently picks up a black-brown mouse that is wearing an electrode on his head, which sends eight tiny wires into his olfactory tubercle, each eavesdropping on a few neurons. The mouse is also fitted with a device that measures when he breathes, so they can correlate his brain activity with each odor's inhalation. The protrusions along his snout make him look like a little furry dinosaur.

Gadziola places the mouse inside a little plastic tunnel, like the kind you might find in a hamster cage, which is itself inside a box into which she can pump odors. She gently attaches a connector to the electrode on his head, and makes sure he is lined up to face the lick spout, a tiny water dispenser, and an odor port, which will release puffs of scented air. The mouse is being trained to discriminate between two scents: banana and pineapple. If he licks the spout when the banana scent comes out, he gets a drop of water. If he licks when the scent's pineapple, no water. As the mouse sniffs, a monitor shows his brain activity, eight lines that spike as his neurons respond. (Carlson is also investigating odor processing in rodents, but she studies rats free to move around a cage instead of holding still, and tracks the larger oscillations in the brain, rather than activity from a few neurons.)

Both are using healthy animals, and they hope to gain fundamental information about the neural signaling of odor detection, discrimination, and adjusting to changes, like when Gadziola swaps the banana "reward" scent for a different one. But, Wesson says, in future studies, they'll use this kind of setup to listen in on the brain activity of Alzheimer's mice, hoping to understand how the disease progresses and how upstream neurons cause problems with information processing downstream before the amyloid has spread. Or perhaps in the future, says Gadziola, if you had a treatment to test, you could record the animal's brain activity before and after applying it, to see if their neural function changes—"Do we actually see it go back to the wild type's reaction?"

All of this, they think, will illustrate the fundamental problem underlying smell loss: neurons handing off the wrong amount of information. Remember, odor recognition is performed by a network of neurons that code a pattern together. "Let's say you need the activation of 500 synapses to know that you smelled a rose," says Wesson. Early in Alzheimer's, when the olfactory bulb becomes hyperactive, perhaps 700 will activate instead. "So you might just not be smelling rose. You might have some cloudiness," he says. Later, when pathogenesis takes over, you have the opposite problem: you no longer have *enough* properly working synapses. If only half of the rose pattern lights up, you can't interpret it.

And that's why discrimination *between* scents becomes impaired, not your overall ability to smell. Recall Gottfried's topographical maps: Kansas is mint and Idaho is lemon. Imagine that as neurons become damaged, the data points on those maps wink out. The distinctions that helped you identify each state grow blurrier. "Let's say you had a map of Arizona and one defining region of that is Sedona National Park. Then suddenly that is flattened because maybe an asteroid hits it," says Gottfried. "Now you look on your computer at the topography and you say, 'That doesn't really look like Arizona anymore.'"

As the maps grow fainter, they also become harder to tell apart. When that happens, Arizona is like New Mexico. Kansas is like Idaho. Mint is like lemon.

FOR YEARS, EXPERTS HAVE BEEN TRYING to develop a smell test as an early warning system for Alzheimer's, to pinpoint the moment when the maps begin to fade. Since olfactory loss comes first, a smell test could catch dysfunction before neurological damage would be visible in an fMRI or amyloid imaging scan, even before the person notices anything is wrong. Doctors who like the idea of a smell test argue that prompt diagnosis means earlier care, and perhaps the delay of impairment. For physicians and drug makers, being able to track Alzheimer's from its first stages could also mean better-tailored treatments, and ones meant for earlier use. But critics argue that so far, smell tests aren't good enough at distinguishing Alzheimer's from more common causes of olfactory loss, and could spook patients unnecessarily.

If you want to see smell tests in action, you have to visit Dr. Richard Doty's clinic at the University of Pennsylvania's Smell and Taste Center. Home-brewed smell tests are an old diagnostic tradition: Many doctors cobble them together by stuffing fragrant items like cinnamon sticks and garlic salt into jars. But Doty, a psychologist and ear, nose, and throat expert who directs and cofounded the center, turned that idea into an olfactory obstacle course that for three days every month is open to nonsmelling members of the public whose perplexed doctors have sent them here as a last resort.

The folks waiting for Doty on this particular clinic day range from a tween boy who has never been able to smell to a woman in her 40s who has only smelled a bad recurring odor since the night she fainted and hit her head on the coffee table. But there is one visitor for whom the diseases of aging could be the culprit. This is Howard, who along with his wife, Esther, has just taken a seat across from Doty's desk. Howard is tall, white haired, and quiet. Esther, a dry wit, does most of the talking for him. Together, they have that wonderful telepathy of the long-married couple used to doing everything in tandem. "Can I just tell you that he's 90 years old and he's still flying a plane solo?" Esther says by way of introduction.

"That's fantastic," says Doty.

"But he can't taste," says Esther. Over the last several months, her husband's lost pleasure in eating. He can taste sweet a bit, but salt seems exaggerated and other seasonings taste bitter. "Most everything is just bland," says Howard.

"How about your smell?" the doctor asks.

"I hadn't noticed it wasn't working good."

The doctor nods. "Often what we see with complaints of taste loss are it's really a smell problem. The taste buds only signify sweet, sour, bitter, and salty. Everything else, like steak sauce, chocolate, coffee—all these things are really smells," he says. They go over Howard's medical history for red flags: recent illnesses, new medications, exposure to chemicals. Nothing stands out. "Over the age of 80, three out of four people have some demonstrable loss of smell. Between 65 and 80, half the population. Among that group, there's about 10 percent that have total loss. So what you're experiencing may simply be a correlate of aging," Doty tells them. "But we'll see."

Howard kindly lets me tail him through an eight-hour-long round-robin of exams, many of them invented by Doty. One involves putting his face into the cavity of what looks like a giant hairdryer diffuser. This is Doty's most recent invention, which tests Howard's smelling sensitivity by blowing different concentrations of a rose oil–like scent at him. It sends up two puffs of air, and Howard's job is to pick which one is strongest. In another, designed to measure whether he truly can't taste, a young researcher pipettes drops of clear fluid—salty, sweet, bitter, or sour—onto Howard's tongue, asking him to name the taste and rate its intensity. At other stations he has his nose probed for obstructions and airflow problems. He holds an electrogustometer, a sort of metal wand, to his tongue and judges which of two mild electrical currents prompts the strongest taste, a way of checking taste bud functionality. He takes a cognitive exam that asks him to repeat lists, spell words backward, copy a drawing, and fold a piece of paper and place it on the floor.

But this clinic's most famous exams are its scratch-and-sniff tests. The first is known fondly as the UPSIT—that's short for the University of Pennsylvania Smell Identification Test. It assesses how well you can distinguish between odors. Doty came up with the brightly colored booklets in the 1980s, a 40-item test asking users to smell common scents—turpentine, licorice, dill pickle—and pick the right name from four choices. The second is the Odor Memory Test, and Howard's is administered by Geraldine Brennan, one of Doty's staffers, in a conference room, all wood paneling and curio cabinets full of vintage medical devices. This time, she explains, instead of trying to identify the scent, he'll just be given one, presented as a scratch-and-sniff tab in a booklet. They'll wait for ten, thirty, or sixty seconds while Howard counts backward from 280 by threes. Then Brennan will turn the page, revealing four more scratch tabs. One will be a repeat of the scent. The other three will be misleads. Howard must choose the correct one.

Brennan uses a pencil to make hash marks on the first scent tab, releasing its odor. After ten seconds, she flips the page and scratches each of the four tabs, offering them to Howard in turn. "Which one of these four do you think best represents that original one?" she asks.

"I have no idea," he replies good-naturedly. But it's a forced-choice test, so he picks one. And the test keeps getting harder, as the delays between the original and follow-up sniffs get longer. All the scents are common odors, but Howard only recognizes a few. There's one he calls "the stink." Then there's "the banana one." Technically, naming the scents is not part of the test—it's just something he's doing to help with recall.

After a dozen rounds, Brennan cheerfully tells him he's all done and can head back to the waiting room after a very long day. "Good! I would venture to say I got 2 out of 40," Howard deadpans.

"Nah, you only had 12," Brennan says. Howard chuckles. "I know," he teases her. "It seems like 40."

His journey through the round-robin illustrates what there is to like about smell tests: They are easy to take; the equipment is simple; the tests are non-invasive and much cheaper than, say, time in an MRI. (The UPSIT costs $27.) The World Health Organization estimates that by the year 2050, more than 70 percent of the people living with dementia worldwide will be in low- and middle-income countries. For clinicians in impoverished areas, it's hard to beat scratch-and-sniff tests with long shelf lives and no moving parts.

Most importantly, advocates of smell tests argue that when used with other exams, they lead to better diagnoses. In 2008, Dr. Davangere Devanand, a psychiatrist and neurologist at Columbia University, and the codirector of the Memory Disorders Center at New York Presbyterian Hospital, published the results of a three-year study in which his group found that the UPSIT, when used with a common memory test, functional ability, and MRI measures, increased the accuracy of predicting which patients with mild cognitive impairment would convert to full-blown Alzheimer's.

"People who have a low score on this [smell] test are four to five times as likely to develop Alzheimer's as another person with memory loss who doesn't score low on the test," says Devanand. Combined with the other indicators, he says, "we are much more accurate in predicting what the outcome will be." Using several tests at once is no different, he says, than weighing a half-dozen risk factors—such as family history, cholesterol level, and obesity—to predict the likelihood of heart attack.

These tests work because they cast a wide net, says Doty. By requiring you to match an odor to its name, the UPSIT checks your powers of detection, identification, discrimination, *and* memory. "It encapsulates all aspects of the function in that system, including the ability to connect meaning to whatever is being smelled," he says. "It is looking at a bigger picture of the function of the system."

But olfactory loss occurs with many diseases, not just other cognitive disorders like Parkinson's, but schizophrenia, depression, and the common cold. Smell tests alone can't diagnose which one you've got. This lack of specificity—and the possibility of false positives among neurologically normal test-takers who just happen to have a crummy sense of smell—means the tests aren't universally embraced. Moulias, the French doctor, points out that they would be better at discriminating between abnormal and healthy aging if you had a record of how well each person smelled when they were younger. And Herz says she suspects that at the earliest stages of disease, smell tests are picking up a *verbal* deficit, not an olfactory one—the ability to connect odor to name.

So now Howard is back in the waiting room, sitting next to Esther, resignedly crunching his way through an apple. "Bland," he complains. As they wait, they talk about how smell loss has diminished his quality of life. Esther asks if her husband could taste the veal marsala during last night's dinner. Howard says it was "about 30 percent," and she sighs sadly. Nobody here has mentioned Alzheimer's, or any of the illnesses that can be behind smell loss for a 90-year-old. But Howard is pretty sure what the options are. "It's one of two things," he says drily. "The one-word answer is 'hopeless.' And the two-word answer is 'old age.'"

"He has a good sense of humor. It might be true, though," his wife says comfortingly. "At least he will know that he's tried."

At that moment, Doty calls them in. There are a few nervous moments as everyone jokes around, and then the doctor offers his assessment: It's old age. Howard's tests give no indications of dementia. "On the Odor Memory Test, where you had to remember and count backwards, that terrible test—you did well on that," says Doty. With the rose oil test, Howard's performance was normal for a person of any age. Overall, he did better than three quarters of people in their 90s. But he does have a moderate loss of sensation, the kind that begins for most people around age 65.

Howard and Esther keep saying "Oh, wow!" as Doty ticks off the results. He recommends things to try: a supplement, a special cookbook, making a home version of a smell kit to practice sniffing. Just as Camilli does at the Atelier Olfactif, Doty advises his patient to keep working his nose.

"Anyway, so you're not hopeless," says Esther, fondly ribbing her husband.

"No, I'm not hopeless," he repeats. He gives his wife a sly look. "In that area, anyhow."

ALIENOR MASSENET'S DESK IS ENTIRELY COVERED in little square glass bottles, in an office that is itself a little glass square with a view overlooking a spring-time park. She is a perfumer at the Parisian office of International Flavors & Fragrances, and she built the scent kit for the Atelier Olfactif. But today she has a different task: cutting it back down.

There are 120 scents in the kit, and she wants to get it to 80, the odors she calls the most "direct"—simple, readily identified, evocative. Creating pure scents is an interesting challenge for a perfumer, who normally layers notes to express an abstract idea that unfurls over time. But such intricacy is too frustrating for someone with a head injury or dementia. So Massenet con-templates the program's "*bibliothèque*," or library, a long, skinny box of bot-tles. She pulls one out and dips in two blotters, handing one to me and waving the other under her nose. It has a sweet, airy scent, but it's hard to place. Massenet, who volunteers with the program's head trauma group, immedi-ately does what Camilli did with the seniors, asking prompting questions: "Is it something that you eat or is it something that you see in a house?"

Maybe citrus? "Something that you eat. It's not a citrus," she says. "Is it a fruit or is it a vegetable?"

Fruit? *Wrong.* Flowers? *Wrong.* Wood? *Wrong.*

"See, this kind of smell is not a direct smell, because you can evoke a lot of stuff," Massenet says with a knowing shake of the head. It's a sweet vege-table, she hints. Carrots? *Wrong.* "Not far. Something that Americans love to eat, especially in fall."

Umm . . .

"And it's the same color as a carrot."

I take a desperate stab. Pumpkin? *Yes.*

Now that's strange. To me, this smells nothing like pumpkin, because the scent Massenet has created is *raw* pumpkin. For pumpkin, I think of the

American holidays Thanksgiving and Halloween, of pie, of cinnamon and nutmeg, of the roasting scent from jack-o'-lanterns. Pumpkin, to me, is *cooked*.

As I say this, Massenet's eyes grow wide and she listens intently, because this is exactly the translation problem she must confront as she tries to create scents that are "direct" for everyone. We all have different memory connections to odors, tempered by where we come from, the foods we grew up eating, the holidays we celebrate. So in France, pumpkin should smell like a vegetable. In the United States, it should smell like a spice.

"This one, it's a tricky one," she says, handing over another blotter. It's sweet, perhaps a tutti frutti. Bubblegum? Cotton candy?

"It's a flavor that is *in* the bubble gum," she hints. After some more sniffing, I get it: melon. And now the scent blooms in a new way under my nose. Once I have a word association, the most melon-like aspects of the scent— the muskiness, the cloying funk—become clear. Just as with taste, having a word has focused the perception. But, cautions Massenet, "when you told me it smells like gum and candy, it's not wrong." It's just that my melon is not Delphine's melon, the scent of fresh fruit and summers in the south of France. I smell the melon of a person who grew up in California in the 1980s, who interacted with it through the medium of Jolly Ranchers and Hubba Bubba.

"I tell you," Massenet says firmly, "it's just a question of culture." And as I sit here sniffing the blotter, I realize that she must contend with a permutation of the language problem, and that both are subsets of the broader problem of attention. We perceive what we have learned to pay attention to—through words, through cultural associations, through personal memories. Even with the same odorant molecules locking into our noses, our perceptions of scents may be wildly different.

The people working on an Alzheimer's smell test have run into this very problem. The UPSIT was built for people familiar with American foods and landscapes. But around the world, "people don't know about licorice; they don't know about skunk," Doty says. "Pumpkin pie, they have never smelled it." So now, researchers are creating versions for other countries. Studies in Taiwan, Australia, and Brazil have found that people do better on the test if it's adjusted for local fragrances and terms. The Brazilian version, for example, swapped out root beer for tire rubber, and changed "pine" to the more generic "wood."

And it's not simply a matter of national background—other divides influence what we think we smell. As an example, Massenet hands me a blotter with a sharp flowery scent: laundry detergent. When she's passed it around in groups, women instantly recognize it, and men don't, because not all of them wash their own clothes. But she says, "If I show them the odor of pet-rol, all the men"—she snaps her fingers—"two seconds, they tell you the smell of it: gas." At Doty's clinic, a patient in the waiting room told me he thought he'd struggled with some UPSIT cues because he was too *young*. The test was developed three decades ago for a generation accustomed to washing with bar soap and baking with cloves—two of the test's choices. Even your consumer history plays a role: Your idea of how shampoo smells could have sprung from a formulation based on roses or apples or *anything*.

But there are a few scents that seem to reliably break through ambiguities, and these Massenet wants to keep in the kit. Coffee. Vanilla. "Try this one," she says. No question: *chocolate*. "See, that's direct." Not all direct smells are nice. Take the scent of blood, which she has actually created. "This you will get automatically," she says, and she's right. It's metallic, instantly repugnant, a square blow to the sinuses. For me, it's the smell of dissection day in biol-ogy class. To someone in the hospital, hooked to a transfusion line or recently out of surgery, it can mean a good deal more.

So Massenet gets back to winnowing her library, pulling out old scents and adding a few she's just created. She calls in two colleagues, Celine Manetta, a research fellow, and Dominique Valentin, an expert on the intersection be-tween psychology and food science, and the three begin a cheery, if casually brutal, assessment of the scents under construction.

They studiously sniff the first candidate, and there is a long pause before Valentin announces it's honey. "I had to concentrate to find it," she says. Mas-senet shakes her head, discouraged. "It's not strong enough," she agrees. Just as with the melon, I had been fishing for the concept—hazelnuts? Baklava?—but once I have the word, the scent resolves itself into the now-obvious smell of honey.

For the next one, a smoky odor, Valentin says, "This is garlic, no? Onion?" Massenet is unhappy with it. Too vague. The smell she had been aiming for, she says, is "very natural garlic, the one that you want to cut and eat. *Real* garlic. But this is not good. *Garbage*."

The next one they literally throw in the garbage—it's the scent of natural gas, or at least of the odorant French utilities add to alert customers of leaks. All three French women instantly recognize this smell, different than the rotten egg odor Americans associate with gas; to me it's an animal waste smell, like a bad day at the circus. "It's ultrapowerful. It's too strong," says Massenet, collecting their blotters into a trash bin and taking it into the hall.

Nothing passes their test. Making things easy turns out to be hard. With odors there are no primaries, just millions of patterns, so it's difficult to create a pattern that is universally recognized. And it's just as hard to be the recognizer. Today I have been chagrined by my repeated failure to put word to scent. Even if there is no "right," I feel somehow adrift.

Massenet gives me one more scent, something that occurs to her as she talks about how hard it is to create the seemingly simple. The odor on the blotter is warm and soft. I think of pizza dough, of floured crust. But the scent is also sweet, with just the faintest traces of almond and vanilla. "This smell, you show it in France, you show it in Italy or even in the U.S., everybody answers differently," she says. And for me, an American with happy memories of sitting at my Italian grandparents' table, it is suddenly quite clear that this is biscotti.

But to Massenet, the French perfumer, it is something else instead. It is the madeleine. *The madeleine!* Proust's hobbyhorse, the famous cookie of lost time, the international literary symbol of remembering! Which, it turns out, *I cannot remember.* I came to France to smell lost memories, and I could not smell *the* lost memory.

As I sit there stewing at my dull nose, I realize this makes sense. I could not make the link because I am not French. I have no aunt from Combray. I did not spend my childhood dipping this biscuit into tea. To me, vanilla and almond and baked dough is something else. And so here is the thing about scent and memory and time and the cultural-linguistic wiring that connects one to the other: We all have Proustian memories, sure. But we don't have *Proust's* memories.

IF YOU SPEAK A DIFFERENT SCENT LANGUAGE, perhaps you need a different *bibliothèque*. And there is, in fact, a second olfactotherapy effort just getting off the ground, rooted in an entirely different culture. It's called the Smell A Memory project, and it's based in Singapore.

The project began after a creative team from advertising company J. Walter Thompson (JWT) Singapore attended a fragrance workshop and found themselves thinking about the bond between scent and memory. Inspired, some of them began to read up on Alzheimer's. "The population is aging worldwide rapidly, and when we looked at the numbers, we were quite shocked to discover the percentage of people who have dementia or Alzheimer's," recalls Ai-lin Tan, a creative director for JWT. They were moved by stories about patients withdrawing from their families. "Imagine that you and your mom were super close, and now you can't even have a proper conversation with her because she can't remember you," says Tan. Maybe, the team thought, creating a broad palette of the smells of Mom's youth would give her a way to reminisce, and a way for you to talk with her again.

Unlike the French program, Smell A Memory is meant only for people with Alzheimer's, and the scents evoke a bygone Singapore, a multicultural community of mixed Chinese, Indian, Malaysian, and Eurasian descent. With Givaudan, the Swiss flavoring and fragrance house, they created a library that would resonate with people who grew up in Southeast Asia.

These come in a clear box containing ten plastic pods, each filled with a small piece of scented foam, selected from a master palette of 63 scents according to the person's ethnicity and cultural background. There is "Hainanese Coffee," the odor of a local brew in which the beans are roasted in butter and sugar. "Fireworks Celebration" recalls the now-banned firecrackers once used for Chinese New Year festivities. There are cooking scents like ginger, chili, tamarind, pandan leaf. Instead of being "direct," some of the scents are atmospheric, like "Bedtime Stories," which mixes the odors of talc and cotton sheets to recall that moment of lying in bed after taking a bath, and "Seaside," which smells of the ocean and sunscreen, and "Opium Nights," which you probably only understand if you were part of a certain place and generation. They tried to re-create the scent of kampongs, or the rural villages in which many elderly Singaporeans grew up, by combining the *je ne sais quoi* of grass, pigs, and chicken droppings, but kept only the grass—it was just too hard to get the mix right.

In 2013, the group rolled out a pilot program at two nursing homes. Instead of prompting people to recognize odors, Tan and her colleagues encouraged them to free associate. "There is no right or wrong answer with a smell," says Tan. "You just want to have a conversation around it." Just as the French volunteers did, the JWT team quickly saw how dementia could rob a person's sense of smell. Tan specifically recalls a bald, gaunt man in his 80s who, despite a few missing teeth and a thin layer of stubble, had a distinguished air—the JWT team wondered if he had been a schoolteacher or an army officer as a young man in his native India. No matter which scent they would pass him, he would just shake his head. He didn't know what they were.

"It doesn't matter," Tan urged. "Whatever you think it is, just tell us."

Still no luck. The team huddled for a moment, tried another scent. This time, the man broke into a smile. "It's roses!" he said authoritatively.

It wasn't.

The scent was actually jasmine, selected for him because it's a beloved fragrance in India, commonly associated with the flower garlands used as temple offerings. But no matter. To him, it was roses, and with that scent came another word: *girlfriend*. From that association, his memories unspooled. As the women urged him on, he told them, a few words at a time, about his teenage years, how he attended a private boys' school, how he and his classmates saved their pocket money to buy roses for the girls at the convent school across the road.

And this is the magic of olfactotherapy for Alzheimer's. If you recall Gott-fried's neural maps with their disappearing topographical points, here was a man whose map collection had faded. To him, jasmine was like rose. But having a bad map doesn't matter if you don't care about going anywhere in particular, if your focus is simply using a recollection—*any* recollection—to provoke a present-day emotional experience. Even though the man got there the wrong way, that memory of giving flowers to long-ago girlfriends still existed, locked in a part of his brain less damaged by dementia. And because scent memory is so evocative, those feelings were still sweet.

Tan recalls that one rainy day, as her colleagues sat around a table in a nursing home dining room waiting for someone stuck in bad-weather traffic, they noticed they were being closely observed by someone who wasn't part of their study. She was a spry woman in her 90s, with close-cropped iron-gray hair and a button nose, who ambled over and plunked herself down at the table.

Tan was instantly charmed. "Oh, auntie, you seem really curious. Do you want to have a little game with us?" she offered. The woman quickly agreed. Thinking they'd offer her a simple but pleasant odor, they wafted the scent of mango under her nose. To their surprise, she recoiled. "Oh, I hate this! It's horrible!" she said. "It smells like *dog urine*."

Think about what's happening in the brain at this moment. Two high, sweet odors connect with memories, one of fruit, one of dogs. Perhaps the olfactory neurons are not making enough electrical handshakes. The data points are off. The wrong scent map is recalled. Mint is like lemon. Jasmine is like rose. Mango is like dog urine.

That's the perception. But here is what is actually happening in the nursing home, because olfactotherapy lets people travel to the past despite their faded maps. Cued by the wrong scent memory, an elderly woman tells a group of eager young listeners stories about growing up in Malaysia. She tells them about getting married and moving to a kampong, about how she had three dogs, about the naughty one that followed her everywhere. She tells them about going to the seaside, and how she loved to dance. She gets up and begins singing a little song, something like "Que Sera, Sera," but she makes up her own words. She holds her hands out to one of the young women, who takes them in her own.

And they begin to dance.

THREE

·····································

Vision

DEAN LLOYD IS JUST OFF THE BUS, standing on the corner, holding his cane and a black leather briefcase, about to head into his office for the day. Lloyd is a broad-shouldered man with wavy salt-and-pepper hair and a firm, booming voice that is part lawyer and part South Dakota ranch farm heritage. From the cane and his heavy black eyewear, a thick set of wraparound plastic frames with darkly tinted lenses, passersby on this Palo Alto street might think he can't see at all. Lloyd was, in fact, once blind. But now, he can see. Or, at least, he can see something.

Lloyd is one of the first people in the world with a retinal implant. A tiny electrode array at the back of his eye is taking information from the video camera mounted in the center of his frames and converting it into electrical signals that his brain can interpret as visual cues. Technically, his device is called the Argus II Retinal Prosthesis System. If you want to be less rigorous about it, you could call it a bionic eye. Lloyd likes to fondly refer to it as the Model T, as in, "I've got the Model T Ford here in my pocket," something he will say while patting the video processor that links the signal from the camera to the antenna coil implanted in his body. What he means is he knows he's testing a first-generation technology, one of the earliest implantable visual neurostimulators on the planet.

Lloyd was born with vision, but lost it as an adult thanks to the creeping damage wrought by retinitis pigmentosa, a genetic condition that destroys the eye's photoreceptor cells, which sense light. He was essentially blind for 17 years, unable to do more than tell day from night. In 2007, he volunteered for the Argus II clinical trial, becoming its seventh recipient in the United States and one of 30 trial participants worldwide. That put Lloyd in the curious position of being one of the few people on the planet who has relearned how to see.

It's easy to think of sight as passive, something that just happens, but it's actually an active, interpretive process. Vision is our dominant sense, but we can't parse every bit of the deluge of visual information on offer from the world, much of it ambiguous, complex, and coming at you incredibly fast. Just as external factors like language and culture shape what we learn to attend to, our sensory systems also have neural mechanisms that help us screen signal from noise. Knowing what to attend to is a huge task that implant recipients like Lloyd must remaster.

Lloyd and his fellow trial participants have been sharing their experiences with Second Sight, the developer of the Argus II, in an effort to improve future versions. (The Sylmar, California–based company began selling the device in the United States in early 2014 and in Europe in late 2011, making theirs the first commercially available retinal implant in the world.) Their experiences make for a fascinating glimpse at how sensory input works. The technical term for this process is "transduction," or converting real-world stimuli into the electrical language spoken by the nerves and brain. Learning how to talk *to* the brain—and later how to read instructions *from* it—is the beginning of a story arc that will take us through Chapters 4 and 5, showing us how researchers are studying the brain's language. Neuroprosthetics like retinal implants deliver information *to* the brain, or they "write in," mimicking the work ordinarily done by the sensory organs. With taste and smell, incoming information starts off as chemical signals, molecules that lock into receptors. With hearing, it begins with pressure changes in the air, otherwise known as sound waves. With touch, information comes from mechanical pressure on the skin. And with vision, the base input is light.

Lloyd remembers light. He remembers colors. He remembers writing and letters. He remembers what objects and people should look like. Because he has images in his memory, he knows that the visual impressions offered to him via the Argus are simple. But they are meaningful, and the proof is how he uses them every day. "The human brain has to work with whatever's it got. It can make sense out of nonsense," says Lloyd. "And it's the most marvelous organ that we have in our body."

It's a warm October morning; the sun is recently up and the light still gloomy. Lloyd begins to cross the street. He sweeps his cane vigorously, making a rattling noise on the asphalt, and briskly narrates what he sees as he walks. He doesn't see images, two- or three-dimensional figures. An apple does not look curvy to him; a bus does not look like a cylinder; its wheels are not round. "But I can see boundaries and borders," he says. By these he means contrast points, edges between dark and light areas. These appear as flashes of light, what he calls "perception points." They help him sense the edges of objects or spaces, and he uses them as navigation waypoints.

"I think there was a white line there by the stop sign," Lloyd says, looking down at a turn lane painted onto the street. The Argus has picked out the contrast between the paint and the darker asphalt, he says, "and I use that as kind of a marker."

He keeps walking, passing a small business center. Lloyd notes the pale gray curb that materializes on his right, observes that it briefly disappears when a driveway cuts through it. He notes the border between the sidewalk underfoot and a nearby planter bed of gravel. "If I look off to the street, there, that's asphalt," he says, indicating the much darker roadway to his left.

He doesn't just do this with the black-and-white grid of city streets. The whole world is rendered to him as bright points that mark his surroundings' most reflective and high-contrast parts. He sees oncoming cars as flashes off their windshields. He sees the windows in buildings as the glints off their glass. He can tell the size of a household object, say, a dinner plate on a table, by timing how long it takes to scan between the flashes on its left edge and the ones on its right. He would recognize the letter E, if you blew it up really big on a computer screen, by the contrast points produced where its four bars meet.

And this is how he does it, minute by minute, day by day. He collects a memory of flashes, none lasting more than a second, and uses them to draw a mental image of the world. The process works best in familiar areas; he combines what he gets from the Argus with his older visual memories and his ongoing kinesthetic ones of walking around town. "I know Palo Alto very well. I used to drive regularly. I have a whole picture of the Bay Area in my brain. When people get lost, they usually just call me." He chuckles for a minute. "I'm your GPS, OK?"

He is not kidding about the GPS thing. Lloyd's ability to navigate is uncanny. He uses the cane outdoors; indoors, he gets around mostly by memory, touch, and Argus. He relies on people for some tasks, like an assistant who reads him legal briefs. He uses a few electronic aids, like a watch that speaks the time. But he has no seeing-eye dog; he does not read Braille; he types on an ordinary typewriter. Partly, that's because he has practiced family law for over 30 years, a detail-intensive discipline that required him to hone an already-prodigious memory. But partly it's because he, like his Argus, can work with very sparse information. "The Argus sees a very small spectrum of light, just white, gray, nothing—which is black, of course," Lloyd says. "There's not much to work with. But it works with that little amount of information to benefit you in the maximum way."

He's now reached his office, a suite in a quiet courtyard building, and headed for the kitchen, where he is loading up the coffee machine. "I can see the edges on the coffee maker," he says, a shiny white cube of plastic on a white countertop against a white wall. All that white's a challenge—a lot of flashing, tough to get a clear contrast point. But it's enough that he can suss the machine's edges, size, and shape, and get the gist: coffeemaker. "I see it as a boundary or border and my brain fills in the empty spots," he says.

Organic material is harder. "Biology doesn't radiate enough separate energy to really make a difference," he says wryly. To him, the rest of us are mostly sound, although sometimes he can glimpse our shiniest parts. "Now I am checking you out, see? I am getting your corners," he says, peering closer as the coffee perks. "Your eyes reflect," he says, waving a hand toward them, and for a moment we are both stumped. Then we get it. He's looking at my glasses.

IF YOU DO NOT HAPPEN TO HAVE AN ARGUS II, this is how your vision system works.

You sense the segment of the electromagnetic radiation band called visible light, the wavelengths between about 400 and 700 nanometers. (The bands we cannot see include X-rays, ultraviolet and infrared light, and radio and television waves.) Light falls through the eye's transparent cornea and the crystalline lens beneath it, which together refract and focus the light. The amount of light entering the lens is metered by the iris, the colored portion of your eye, which controls the size of the pupil, the hole at its center. In low light, the pupil expands; in bright light, it contracts. The light passes through the vitreous humor, the clear gel-filled interior of the eye, until the now-focused image strikes the retina, the neural area at the back where transduction, or writing in, begins.

There are two kinds of photoreceptor cells here: rods and cones. Rods work best in low light, and cones in bright light. Cones also give us our perception of color. And it's important to note that color is a perception because— mind-blowingly—*objects in the world have no inherent color.* What we perceive as color is really the wavelengths of light that are reflected from surfaces, with the shorter ones registering to us as violet, and the longest wavelengths as red. The interaction between those wavelengths and our cones creates the experience of color, not anything native to the object or the wavelength itself. (That's also why color "disappears" in the dark, when your rods are at their best but your cones are not.)

You might have learned in school that there are three kinds of cones—blue, green, and red—but the truth is a little trickier. It's more accurate to say that cones respond to all wavelengths but are tuned to be more sensitive to short, medium, or long ones. The pattern of the combined response from the three cone types produces the perception of color.

Early stages of visual processing begin in the retina, where your visual system applies the first of many filters that gate information from the outside world. Ganglion cells, which receive information passed up from the photo-receptors, are a good example. There are two kinds. ON-center cells are sensitive to light that falls in the center of their receptive fields; they increase their neural firing rate when light is detected there, and slow their firing when light falls on their outer edge, called the surround. OFF-center cells do the opposite. Both are also sensitive to the contrast between the light intensity in their center and the surround; they can sense the degree of difference between them. And ganglion cells care more about this contrast than they do about relaying the absolute intensity of the light around you—whether the room is illuminated by stadium lighting or a single match. They are tuned to picking up *differences* in brightness between two specific areas in the visual field, a contrast that will remain steady regardless of ambient lighting. So here your visual system has made a choice for you: It's going to pass along information about contrast at the expense of information about overall brightness, because that's more helpful in deciphering what you are seeing.

This kind of filtering is repeated many times as information works its way through the visual pathway, and neurons tuned to different features, or aspects, of the incoming information preferentially pass its most salient details along. (We'll go deeper into the neuroscience behind this filtering system in Chapter 4; it is an extraordinarily complex study, which neuroscientists are just beginning to unpack.) For the moment, the important thing to know is that by the time the information has left the eye via the optic nerve, which is formed by the combined axons of the ganglion cells, this processing has just begun. As information passes through the visual cortex, neurons tuned for qualities like orientation, size, and direction of movement parse this information until we ultimately perceive an image. It's important to remember that this is an *image*—not an objective rendering of reality. It's what your brain has constructed from the available information, using its highly biased filtering system and the narrow spectrum of the electromagnetic band human eyes can read. Your visual world may have a finer grain of information than Dean Lloyd's, because you have some 100 million working photoreceptors and he has very few, but it's just as constructed.

The Argus II cuts in right before this processing begins, intercepting light before it hits the retina. And it might go without saying, but this is a very, very hard form of biohacking. Like the complementary read-out techniques we'll explore in Chapters 4 and 5, technologies that work directly with the brain or the sensory organs are highly technical, deeply invasive, and largely under the purview of university and clinical researchers. A few, like the Argus, are for sale, but to only a very small client pool. All of them, so far, were developed as assistive devices for people with medical needs, to restore sensory function to those either who were born without it or who lost it as a result of illness or accident. While some futurists we'll meet later in this book dream of write-in implants that could heighten or expand sensory experiences for otherwise healthy people, today's neuroprosthetics are much simpler. They turn the flow of incoming data into a minimalist sketch, not a psychedelic sound-and-light show.

People who have retinitis pigmentosa were perfect to be the Argus' clinical testers, says Second Sight VP of Business Development Brian Mech, because while it destroys rods and cones, it leaves behind working ganglion and bipolar cells, plus a healthy optic nerve. Since some cells still work, you can stimulate them using an implant. "What we do is we fire those cells, and then we rely on the natural visual pathway to take over for the rest of the way," he says.

Argus II recipients no longer use the optics of their own eyes. Instead, a camera mounted on the bridge of their glasses captures and manipulates the light that is passed to their retinas. But first, the camera feeds that information down a cable into a video processing unit that users carry in their pocket. The processor can enhance images before sending them back up the cable to a primary disc antenna on the side of their plastic glasses, close to the eye. The antenna beams power and images to the implant inside.

When Lloyd received his implant, the surgeon installed three connected parts: the coil, the "electronics package," and the array. The electronics package is a silver metal puck about the size of an aspirin. This tiny cannister holds the chip that controls how the implant's electrode array fires. The surgeon attached this to the sclera, or white outer tissue, of Lloyd's right eyeball, anchoring it under the conjunctiva, the protective membrane that covers it. This part of the implant is positioned back inside the eye socket; if Lloyd lifts his glasses to show you where the surgery was, you can't see it at all. The surgeon also inserted a secondary antenna, a flat, gold, oval-shaped coil encased in a polymer coating. It receives the signal from the external antenna, passing power and video images to the electronics package. It's also positioned back inside the eye socket, attached to the sclera and covered by the conjunctiva.

Finally, the surgeon positioned the array, the part that makes the communication bridge between machine and flesh. The array begins with a flexible band of silicon polymer tape, which runs from the silver disk of the electronics package across the surface of Lloyd's eye—invisible to the outside world, because it's also under the conjunctiva—and then pierces through the eye's surface near the iris. It passes through the fluid interior of the eye, which under the surgeon's lens is a glittering, aqueous cavern. Its tail end widens into a larger tab, a grid of 60 tiny metallic spots. Each spot is the terminus of a photolithographically patterned wire printed onto the polymer tape. These are the electrodes that will conduct visual stimulation to Lloyd's surviving cells.

The array is slightly curved, designed to lie against the sphere of the eye. The surgeon carefully positioned it over the macula, the central part of the retina, which lies at the very back of the eye's interior. It's here that you can see the damage done to Lloyd's eye by the disease. The tissue is a mottled dark batik caused by changes in the retina's pigmentary layer. The surgeon painstakingly situated the array, and then used a tiny tack to anchor it in place. Now the video information from the Argus camera will make the electrical jump from array to cell body, and then on to Lloyd's optic nerve, which will carry the signal up the normal pathway to the brain.

With these electrodes, Lloyd's retina receives 60 points of information. (Really, for Lloyd it's more like 52; not all of his are working.) You can think of the electrodes as being like the lights on a football scoreboard, each illuminating when they get the signal to turn on. But on Lloyd's scoreboard, the lights activate only briefly, and all the ones needed to illustrate a figure might not illuminate at the same time. It would be hard to use this scoreboard to create persistent images, or ones with fine resolution. There is no detail, not much shading. But it can still convey useful information about objects: where they are, their size, their orientation, their brightness, their absence or presence. If you are willing to learn, you can use it to get around.

The Argus is the direct descendant of a much earlier sensory prosthetic— the cochlear implant. These restore some hearing perception to the deaf by combining an external microphone and sound processor with an internally implanted electrode array that converts acoustic sound into electrical signals, passing them to the auditory nerve. Researchers began experimenting with them in the 1960s and '70s; in the United States, the first multielectrode device was approved for sale in 1984.

Cochlear implants have even fewer inputs than retinal ones, so the entire world of sound must be radically simplified. Some wearers have compared them to pianos with only a few keys, and said that at first the sensation is disorienting and unpleasant. As with the retinal implant, users must learn to sort signal from literal noise, guided by any sound memories they may have. The implants have been criticized over concerns about them being touted as a "cure" for deafness, the variability of outcomes among patients, and the effects of implantation in young children. Yet overall, the industry has been a successful one. By 2012, the federal Food and Drug Administration esti-mated that about 324,000 people worldwide were wearing cochlear implants. These devices paved the way for retinal implants by showing that neuropros-thetics could be worn long term without degrading or causing infection, and that users could adapt to the writing in of very sparse information.

One of Second Sight's cofounders was Alfred Mann, who had previously founded Advanced Bionics, a Southern California manufacturer of cochlear implants. Inspired by a colleague who had retinitis pigmentosa, he'd begun to wonder if a similar prosthetic would work for vision. In the late 1990s, Mann began exploring this idea with several researchers, including Dr. Rob Greenberg—previously at the Johns Hopkins School of Medicine—who had been experimenting with electrically stimulating the retina by placing small probes within the eye. In these experiments, says Mech, patients "could see a spot of light when there was one wire and they could see two spots of light when there were two." That was important—at this early stage, researchers worried an electrical stimulus might produce a useless sheet of light, rather than discrete points. But spots of light give you something to work with. "And so in Greenberg's mind at that point," Mech continues, "the rest of it is just engineering."

Second Sight was founded in 1998, and Greenberg became its CEO and president. The Argus has been in development ever since. The first version was fairly crude, designed to answer two questions, Mech says: Could you safely stimulate the retina over long time periods, and would the effect eventually wear out? Their first implant, the Argus I, was literally built on Advanced Bi-onics technology. "We took an already-approved cochlear implant and just modified the electrode lead on that implant so that it could be used inside the eye," says Mech.

Between 2002 and 2004, a half-dozen clinical trial participants were im-planted with the Argus I. Mech says the results were promising, but limited. The small array—just 16 electrodes, because that's what the cochlear implant had—didn't provide much resolution. And it can be hard to measure improve-ments among people who are nearly blind. You can't use eye charts and tests designed for people with more vision. "All we can do is measure what they can do," Mech says. "They were able to ambulate and avoid objects in their path, find doors and windows. One of the fellows, in his own home, could find countertop appliances like the blender and the coffeemaker and stuff like that. He could go out and identify the roof line and find people in the room."

The next version boosted the number of electrodes while making the electronics package smaller, which in turn brought down the surgery time. The Argus I took eight hours and three surgical specialists to install. The company wanted the next version to be something a single surgeon could do much faster, making the process less risky and, from an insurance standpoint, less expensive. They also had some technical challenges to iron out. One was to make sure the array wouldn't damage the fragile retina, which Mech describes as "like wet tissue paper." For this, they developed its curved design, as well as the polymer mix for the array. The other challenge was finding a way to get the feed-through for 60 electrodes into a tiny, watertight capsule that could sit within the eye. But by 2006, they had their new design.

Now they needed someone who was willing to try it.

IN DEAN LLOYD'S OFFICE, the computer points *away* from his desk chair; it's set up so that his assistants can read the screen for him. But otherwise it's just another law office: bar exam certificate on the wall, files stacked on the desk, scattered coffee mugs. Lloyd is headed for court today, so he's wearing a dark blue pinstriped suit. As he sits behind his desk, his head points slightly up; he hardly ever levels the full barrel of his camera gaze at the people sitting across from him.

He takes off the frames to show his right eye, the one with the implant. You can see the dark ring of his pupil and the hazel green outer disk of his iris, but the electrode array at the back isn't visible, even if you squint rudely into the depths of his eye. The only clue that he's got a neurostimulator is that when he removes his glasses the apparatus starts making a frustrated beeping, a signal that the antenna in the frames has lost linkage with the coil in his eye.

Lloyd had originally planned a career in medicine. "It was kind of like a family tradition, because my grandfather was an old country doctor, and my uncle was, and we just thought I should be one too, I guess," he says. (After three weeks working for a cattle company, he'd also ruled out the other family business: ranching. "The horse hated me and I hated the horse," he says. "We had a mutual disadmiration.") He began medical school at the University of South Dakota, but midway through, started having trouble making out histological samples under a microscope. That puzzled Lloyd, because in ordinary life, his vision was not too bad—maybe 20/40, he says. "I drove a car; I did everything normally," he says. It was only under the microscope that the world looked fuzzy.

He went to an ophthalmologist who misdiagnosed him with Usher syndrome, a condition in which one loses both sight and hearing. Believing that he would be blind in three to five years, Lloyd withdrew from medical school. "They gave me a booby prize in biochemistry," he chuckles, "because I had straight A's in biochemistry and anatomy."

Lloyd worked as a biochemist for the next five years, moving to California and taking a job at Stanford University. While he was there, he got a second medical opinion, and was correctly diagnosed with retinitis pigmentosa. The condition has dozens of genetic variations, and because it varies so widely, the doctor told Lloyd it was impossible to predict the arrival or severity of his symptoms. So the doctor told him not to worry about it, Lloyd recalls: "Just take one day at a time. That's what I did."

After discovering he had an affinity for computers, he switched to software engineering. He got married, had two children, and worked for a variety of Bay Area companies, designing streetlight systems to control traffic flow, medical data software, and a program to help hardware stores mix paint colors. Working with computers was a visually intensive job, with hours of staring at tiny letters on an illuminated screen. Yet by the early 1970s, the promised blindness still hadn't developed. "At nighttime, I didn't think my eyes adapted to darkness as well as other people's eyes did," Lloyd recalls, but otherwise he had no complaints. And then one day, while drilling through some fiberglass to make a backyard patio cover, he got a fragment stuck in his eye and made an emergency trip to the doctor. "He looked in there and said, 'Wow, you've got more than just fiberglass, now,'" Lloyd recalls. "You've got a real mess going on in there. You've got cataracts."

Lloyd, then in his thirties, was awfully young for cataracts, but they can be associated with retinitis pigmentosa. He had surgeries, but they produced only temporary respite, and his loss of acuity became noticeable. "Like, if you look down the street and look for somebody's license plate, you'd find out you couldn't read it," Lloyd recalls. Now his vision faded fast: "Pretty much normal vision to almost no vision in six months." He recalls this in a matter-of-fact tone, almost a victorious one. He'd expected to go blind in perhaps three years, but he'd made it to ten.

As his vision failed, Lloyd had come to rely on the memorization skills he'd used as an anatomy and biochemistry student. He'd memorize code and debug it mentally, touch-typing his corrections. But eventually that wasn't enough, and he had to stop working as a programmer. His marriage ended, too. During child custody negotiations, Lloyd became interested in law. He started taking classes through an evening program at a small college. "I never planned on ever practicing law," he recalls. "I thought, 'Well, I'll use it for my own use.'"

But law, despite its required oceans of reading, turned out to be a good match. A lot of legal practice is oral: Lloyd liked public speaking, felt comfortable in the courtroom, enjoyed the conversational aspect of getting your argument across. He took his law books to a recording center for the blind, where a person read them onto tape. When it came time to take the bar, in which students sitting for the exam have to make their way through reams of reading material, he had two readers speak the texts aloud, both of them sucking on ice and lemons to keep from losing their voices.

Lloyd taught himself how to get around in the courtroom, going in before trials to learn the layout, so he's always facing the right direction and knows where to address the jury. When walking, he'll gently place a hand on a friend's right shoulder and follow a half-step behind. Indoors, he'll sometimes trace his fingers along the wall to get his bearings. Lloyd Lodge, his name for the hillside house he shares with four other business professionals, wraps around a central staircase; he routinely uses that instead of the elevator, counting steps and lightly touching the wall. Space, sound, touch, memory—these were his tools in the 17 years before the Argus. "Necessity is the mother of invention, and you're basically taking those other senses to fill in for what you don't have," he says.

But he also liked the idea of having some of his vision back. He first heard about the Argus II through ophthalmologist Jacque Duncan, MD, who practices at UC San Francisco, one of several test sites for the device. He met the requirements for the trial: He was over fifty. He could tell day from night, and see a sharp point of light shining in his eye, although no more than that. The device was novel, and surgery always carries a risk, but he decided to go for it. "I figured, ultimately, I probably have very little to lose," he says. He became U.S. test subject number 7 in July 2007.

Lloyd is interrupted by a knock on the door; his legal assistant Alex Sandoval is here to go over paperwork for this afternoon's hearing. Sandoval opens up a file containing the details of a child support battle, and reads out a litany of dates and figures. Lloyd listens closely, repeating the numbers. The whole time, he's drawing on a sheet of paper, making notes that sometimes look like writing but mostly look like circles. "You see me scribbling over here? I'm converting oral into memory," he says—the old muscle memory of putting pen to paper helps him remember what he's hearing.

When it's time to go, he loads up his briefcase and then heads over to the bookshelf where two chargers are juicing backup Argus batteries. The morning's battery is almost drained. "Put that little baby in there, start it back up and it'll go for another eight hours," he says, sliding a new one into the video processor unit, slipping it back into his coat pocket, and heading out to Sandoval's car. (On the way to court, I take a separate car and get lost. I have to call Lloyd twice for directions, which he promptly reads out from memory when I give him my intersection. He really was not kidding about the GPS thing.)

At the courthouse, Lloyd, his assistant, and his client huddle in the hallway, and then head inside for what turns out to be a long, dull afternoon. The court is backed up. The judge finally gives Lloyd's client the last docket position of the day, and now Lloyd has to cram all his stats and dates into 20 minutes. As he sits next to his client, Lloyd sometimes leans back in his chair and strokes his fingertips under his chin, as though brushing sugar off of it. Other times, his expression perfectly neutral, he adds to his drawing, now a mass of blue and black circles. (Sandoval is taking the official notes.) When Lloyd speaks, his voice is gravelly, with a slight drawl; he tilts his head a little to the left to address the judge. Every now and then, the Argus beeps softly.

It's one of those frustrating days in court where the only verdict is that everyone will have to come back and do it again. Lloyd and his client head back into the hallway, confer again. It's around this point that it's becoming obvious that even without the retinal implant, Lloyd is also personally kind of a machine. It's been nearly 12 hours since he got on the bus this morning, but he has not stopped for so much as a stretch break. He has changed batteries, but he has not had lunch. Doesn't he need to, you know, eat sometimes? "I had coffee," he says with a shrug, and heads back to the office.

THE SECOND SIGHT HEADQUARTERS IN SYLMAR, an industrial suburb where the I-5 touches down after winding through the scrub of the Los Angeles National Forest, is an almost featureless one-story facility in a bank of nearly identical ones. Inside, Brian Mech is peering through a window into a clean room, where workers are assembling arrays for the Argus II. Under yellow light, necessary for photolithography, people in bunny suits, paper face guards, and blue latex gloves are inspecting the tiny tabs. Mech points at people who are leaning into what look like very fancy microscopes—they're checking to make sure each electrode is working. Farther down the line is a machine that's combining the array with the electronics package, the part that contains the chip. "That turned out to be one of the hardest things, from a technology perspective," Mech says. "This is the smallest, highest-density neurostimulator ever developed and approved for an indication. So putting that many electrical contacts together in such a small space, and yet making sure that they are leak-tight, that no water can short them out, is very hard."

He moves down the hallway, peeks into a second clean room. Here, the electronics packages are laser-welded together, tested for leaks, and inspected again. Soon, they'll be ready to go out into the world. It's the day before Second Sight will announce that it's successfully completed its first round of commercial implants, done at the University of Michigan. Mech can't say how many users they might serve in the near future, but the company estimates that there are 1.5 million people worldwide with retinitis pigmentosa, so you do the math.

They've got competitors. Several research groups at major universities are developing retinal implants, and in 2013, the German company Retina Implant AG received approval to begin European sales for the Alpha IMS, which was also clinically tested among people with retinitis pigmentosa. In this design, a chip with 1,500 light-sensitive photodiodes fitted to electrodes is implanted beneath the retina—rather than in front of it—to stimulate the surviving photoreceptor cells. As with the Argus, the optic nerve then picks up the signals and conveys them as usual to the brain. But since the light-to-signal conversion happens inside the eye, the person does not wear an external camera. (They do have to carry an external power supply, which attaches magnetically to a secondary coil implanted under the scalp behind the ear.)

Other people are proposing alternatives that would avoid implanting a device in the eye. Dr. Sheila Nirenberg's lab at Weill Cornell Medical College envisions a system that would combine a device built into glasses and a technology called optogenetics, which is becoming increasingly important as scientists learn how to write information into the brain. Optogenetics is a way of very precisely controlling neurons by stimulating them with distinct light wavelengths. Optogenetics requires gene therapy to insert opsins, or light-sensitive proteins, into the cells. The genes for the opsins are loaded into the shell of an emptied-out virus, and in this case would be injected into the retinal ganglion cells of the eye. These cells, as you may recall, lie between the photoreceptors and the optic nerve, and they survive even when retinitis pigmentosa and other diseases attack the photoreceptors. So after gene therapy, a person would have a bank of still-healthy cells that can be activated by a very specific kind of light.

Now you need to give them some information to process. Nirenberg, a computational neuroscientist, has argued that the kind of stimulation offered by current retinal prosthetics isn't close enough to normal input, and that's why users are only able to see low-grade information like flashes and edges. If you could more fluently translate real-world images into the brain's electrical language, she believes, you could produce higher-resolution vision. So her lab created equations relating the activity pattern of retinal cells to the light patterns they sense when viewing everyday images. The lab can then reverse this mathematical code to turn images into electronic pulses that re-create the retina's language.

So imagine a blind person trying a treatment that goes something like this: They have the opsins injected into their eye. Then they wear a pair of camera-equipped glasses. Inside the glasses is an "encoder," or a chip that turns the video images into those electronic pulses. A tiny projector inside the glasses converts the electronic signal into a light pattern and beams it at the eye. The ganglion cells, now sensitive to this kind of light, receive a signal that is very close to the pattern the retina would naturally produce, and send it to the optic nerve. Using a mouse-version model of this prosthetic system—which does not include glasses, sorry—Nirenberg's lab showed that you can create retinal activity in blind mice that is close to normal. (She promptly won a MacArthur "Genius Grant.")

But Second Sight has beaten everyone to market. Mech expects that most of their new users will experience better visual acuity than Lloyd, who has what Mech calls low persistence, meaning the visual stimulus doesn't last long—that's why he just sees flashes. Some patients with better persistence say they can see silhouettes, but not three-dimensional images. Some can see a grayscale spectrum, probably between five and seven shades—that's enough to distinguish between, say, a person's face and clothing. This is important, Mech says, because even if users can't see people well enough to identify them, they can gather socially useful information: if people are paying attention or looking elsewhere, when they arrive or move away.

To demonstrate what better performance looks like, Mech pulls up a video of a French trial participant. The man is standing on a plaza, pointing at a woman in a black coat as she passes in front of him, showing that he can track her movement. In another clip, he counts the posts lining a street. And in another, he's walking down the sidewalk, zig-zagging his cane. "Watch what happens here," Mech says excitedly. On-screen, a passerby unexpectedly steps in front of the Argus wearer, who freezes. "He stops. He stops!" says Mech. "Not because he touched him with the cane, but *because he saw he was there.*"

While the data from the Argus is sparse, it can do some neat tricks. On the video processor each user carries, there are three buttons the company can program to give them special powers, abilities the ordinary eye does not have. One of Lloyd's inverts white and black, to make it easier to find doors and windows. There's an enhanced contrast option, to make dark things darker and light things lighter. Another is edge detection, which helps when navigating man-made and indoor environments composed of sharp lines and right angles.

In the future, Mech has suggested, facial recognition software might be helpful. Or, if you want to get a little farther out, he says, "If we connected an infrared camera instead of a regular wavelength camera to the system, you could argue that these patients would have better vision at night than you or I." And there's really no reason that the person would have to use the camera mounted in their frames; you could hook them up to *any* camera, or even an Internet feed. "They could be seeing things from *your* camera on *your* lap-top if that's what they wanted," Mech says, basically giving them true tele-vision—the ability to see at great distance. Currently, the video feed flows by in real time, but isn't stored anywhere. So, I ask, no rewind button? "No," Mech says thoughtfully. "Although that's interesting."

The company has developed a take-home kit to help people relearn what's visually useful. It includes a white magnetic board to which you can stick black shapes and letters, so you can trace them with your hands while looking at them, calibrating visual and touch perception. (Lloyd practices at home with an early version, posterboards embossed with fuzzy geometric shapes.) But recalling what these images mean can be difficult for people who have been blind for a very long time. Mech recalls a clinical trial participant who went through a similar training on computers. The woman "was able to literally trace the letter S on the screen—she could trace the shape; she could clearly see it," he says. "And she didn't know what letter it was, because she'd forgotten."

One of the most intriguing results of the Argus II trials is how varied this relearning process can be. For more than half of the users, Mech says, "it does take them time to relearn and recalibrate their vision. And for the rest it doesn't. It's almost instantaneous. They turn the device on and they just get it. So very often, the people in the first group, they will say, 'Ah, you know, it was so frustrating, so frustrating! I'd do my homework every day, I would use it for two or three hours and I just couldn't understand anything. *Anything!* And then one day, everything started making sense.'"

Mech and his colleagues would really like to know why there's such a range. They estimate at least 100 variables might affect performance, but with only just over 100 Argus users so far—that's including the two clinical trials and the earliest commercial users—there's just not enough data to draw conclu-sions. Even what seems like the most obvious factor, time spent blind, doesn't seem to correlate with anything yet. And, of course, for a disease with so many genetic variations, there may be many answers.

Those answers may become clearer once there are more people with reti-nal implants. So far, the Argus has only been used for those with retinitis pig-mentosa, but it could plausibly be used by people with other conditions that leave some functioning retinal cells, notably age-related macular degenera-tion, which generally destroys central vision but leaves people with some pe-ripheral vision. That would open up a much larger patient pool; the company estimates that worldwide, there are about 2 million people who are legally blind from this disease. But there are even larger populations with other dis-eases that destroy retinas and optic nerves. If you could insert the implant farther along the visual pathway, you might be able to help them, too.

Second Sight is now exploring two ideas for its next-generation implant. One possibility is a higher-grade version that will still go in the eye's interior. The company is pursuing a technology called current steering, which shapes the electrical field between two electrodes on the array to focus on a piece of tissue that lies between them. This creates additional "virtual electrodes," allowing a greater number of inputs using the same device. Along with improvements to the video processor, it would theoretically allow for higher-resolution vision.

But there's probably a biological limit on how many electrodes they can usefully employ, Mech says. There's not a one-to-one match between an electrode and the cell it stimulates. In fact, each might stimulate several. Too many inputs could cause a signal bleed over many cells—a "cloud effect," or a blurry light instead of a pinpoint. (Optogenetics approaches are trying to get around this problem as well.) So Second Sight is also considering developing a cortical prosthesis, which would skip the eye and optic nerve altogether. In this version, the electrode array would rest on the visual cortex and the electronics package would likely be implanted on or in the skull. By jumping the optic nerve and going straight to the brain, Mech says, their implant could be used for nearly any form of otherwise untreatable blindness.

In design, Mech says, "it wouldn't look very different from the Argus II. It would have probably 60 electrodes." A glasses-free version might put the exterior antenna as well as the camera in a special hat, or the camera could be handheld. But this would require some serious technological development. Getting from noise to signal will be even harder, because by skipping over the interpretive work done by the eye and the optic nerve, you're patching your signal in much farther downstream. "You've lost all of that encoding that goes on in the visual pathway before it gets to the cortex. So now you're giving the cortex very raw signals," says Mech. "You'll have to do a lot of learning in the clinic on how to shape those signals and make it so that the brain can understand what is coming in to it, and people can make sense of it."

While it's an astounding proposition, Second Sight is not the only group interested in directly stimulating the brain. The idea is decades old, although early work was stymied by the risk of infection and the difficulty of powering an implanted device. Some of the first successful brain implants were done for motor, rather than sensory, purposes, including deep brain stimulation electrodes for the treatment of muscle tremors in Parkinson's disease. And as we'll see in the touch chapter, Chapter 5, brain implants that would drive prosthetic limbs are under development as well.

In Chapters 4 and 5, we'll dive deeper into how scientists are learning to talk to—and read out from—the sensory areas of the brain, and the kind of thorny questions and fantastical possibilities this opens up. But for now, the Second Sight factory floor is convincing proof that simply writing in to the brain is no longer a sci-fi dream; it's a product. On the way out of the office, we pass through a floodplain of empty cubicles designated to become the sales department. This is a company that clearly intends to grow.

DEAN LLOYD IS RIDING IN THE CAR on the way to his ophthalmology checkup. He goes every six months; part of being a clinical trial participant is periodic exams to make sure the device isn't causing trouble. Even though he's not headed into his law office today, Lloyd's wearing a suit. The only indication that it's a slightly more casual afternoon is that he's wearing his beloved black cowboy boots.

It's a dry winter, and the hills that stretch along the freeway are the same dull gray-brown as the road. Lloyd cranes his head, sees only a flash off the windshield. "I'm not even sure if it will go through the glass," he says of the Argus.

Lloyd has been an avid experimenter with his device, which he wears from sunup till bedtime. One of the most intriguing aspects of vision became readily apparent to him: It requires motion. The eye is constantly moving. We make several fast shifts called saccades every second, which change the eye's fixation point and help us examine a scene in detail. Because Argus users are wearing a camera mounted just above the nose, they can't replicate this by moving their eyes. Instead, they have to scan their entire heads, as birds do. "If you are a farm person, like I am, I can remember the chickens jerking their head back and forth and I wondered why they did it," Lloyd says. "And I didn't realize that so intensely until I had the Argus II implanted."

In fact, his first memory of actually getting any kind of visual impression from the device required learning to do what he calls the "chicken movement." A common misunderstanding about the Argus is that users can see right after their surgeries. But their implants aren't immediately connected to the camera-and-glasses parts of the device. After recovering from their surgery, users have their implants personally tuned. The stimulation threshold for each electrode is measured and set to a frequency range that works best for them. (If the frequency is too slow the image will pulse; too fast and the image will fade.)

They also have to practice with the device before they can take it home, and that involves learning to scan. About a month after his surgery, Lloyd recalls a neurophysiologist taking him on a walk through a garden behind UC San Francisco, stopping before an object, and asking him to identify what he could see. "When he says 'see,' I'm looking for a picture," Lloyd recalls.

"Well, there is no such thing with the Argus device. So [the doctor] says, 'Dean, you've got to move it left and right. I told you about the chicken!'" So Lloyd scanned his head side to side, and then up and down, and put together his first visual impression in 17 years, flashes that told him the object in front of him was about three feet wide and nine feet tall. That was correct, but it wasn't enough to identify it: a statue of one of the university's founders.

With such scanty information, Lloyd wasn't convinced the Argus would be very useful. So he devised what people around him now know as the Sock Test. Lloyd is a gym regular, and he wears white athletic socks to work out in the morning. Before going into court he changes into "something more somber," or black socks. "Well, one day I got them mixed up. I had black on one foot and white on the other," he says. An opposing attorney noticed and razzed him about it: "Hey Lloyd, how are you going to expect to win your case? You haven't even got your socks right!" Stung by this, Lloyd recalls, later at home, "I pull out my stockings and say, 'This turkey attorney gave me a hard time. Let's see if I can sort out the stockings after the wash.'"

The Sock Test involves 30 socks; ten white, gray, and black. On the first test, Lloyd got the white ones but mixed up the rest. But there was an unforeseen variable influencing his experiment. His implant had been tacked too far from his retina, and wasn't making good contact. So he had a second, shorter surgery he and Mech refer to as "the reset," adding a second tack to more securely fasten the array to the back of his eye. After he'd healed, Lloyd tried the Sock Test again and nailed it. "And that convinced me the device really had potential," he says.

As he got better at seeing, he noticed something unexpected: Sometimes the flashes are bursts of color, even though his processor doesn't actually interpret color. Argus users are only expected to see in white, black, and shades of gray—or in some cases yellow, Mech says. The colors Lloyd perceives don't match the real-world objects he's viewing. "If I'm looking at a green tree, it could be pink, purple, orange, red—some ridiculous color," Lloyd says. But he's certain that's what he perceives. "I grew up seeing color. So the recognition factor in the brain is still there," he says.

Mech and Lloyd's doctor think he's right; other users have reported color as well. It could be that an electrode is randomly stimulating a surviving bipolar cell that used to communicate color information from a cone cell, which is why the flash has nothing to do with the color of the object they are viewing. Or perhaps a certain stimulation frequency is producing a mental perception of color. (If that's the case, being able to more accurately tune the frequencies might allow for color versions of the system in the future.) Either way, Lloyd loves these fleeting moments, and finds them beautiful. "Generally, it's red," he says. "Sometimes I will see perfect ruby red, a pretty ruby red. And the blues are interesting. A pure blue would be a luminescent blue, like the sky." Color, to Lloyd, is an important part the world, something he misses. He knows he may never reclaim his ability to see the full spectrum, with all of its subtleties and shadings, but he does have a simpler wish: "Just to look at a green tree and see that it's green."

Through practice, Lloyd has combined these perceptual clues with enough other adaptations that every now and then, someone jokingly accuses him of secretly having normal vision. For example, although Lloyd no longer drives, he has a great feel for the road. In his garage is his "college dream:" a cherry condition 1969 Mustang convertible, forest green with a cream drop-top, better known as the Green Machine. He goes out cruising in it with his friend Bicycle Bob behind the wheel. (Bicycle Bob is so named because they also go out on tandem bike excursions, riding out to the beach or the 49ers football stadium.) Driving the Green Machine, he says lovingly, "would warm all the corners of your heart. It's just so assertive. You step on the accelerator and it literally jumps." And as we roll steadily toward his doctor's office in my much less glamorous battered Saturn, 20 or so minutes into our drive, Lloyd—who, remember, cannot perceive anything beyond the windshield glass—casually mentions that we are passing the exit to Daly City, a nearby suburb. And indeed we are. How did he do that? "Time is distance," he says. "And you're traveling probably about eh, 65, maybe 70, every now and then. And I can just ascertain that from the feel of your car and ascertain that it's moving at a certain speed. And then I measure the time and I come out to the exact spot we should be at."

In general, Lloyd has embraced his tester status; he good-naturedly refers to himself as a guinea pig, he sits quietly through lengthy exams, he enthusiastically testified in favor of the Argus' worthiness to the FDA, and by the time I got to him, he had already given 65 interviews to curious reporters (he's counting). The one thing he will not do is refer to himself as a "patient." ("Because I'm not patient at all. I'm very demanding of myself and of other people around me, OK?") He sees his role as helping Second Sight make a better product for the users who will come after him. "I know this is the Model T Ford. We're not even getting close to the Cadillac," he says.

He is not the least bit fazed by the cyborg nature of his existence; he's calmly accepted that the machine is part of his body, and that both have to be regularly maintained. As we pull up at the hospital and he gets out of the car, he checks to make sure he's got his briefcase, noting that his backup batteries are in there. "If my battery dies," he says mildly, "then I die."

Inside the hospital's darkened examining room, an ophthalmic technician runs a few basic exams: intraocular pressure, blood pressure—all normal. She tests Lloyd's muscles by asking him to move his eyes; she measures and dilates his pupils. Then Dr. Duncan, the ophthalmologist who got him started on this journey, comes into the room and greets him warmly. "Six and a half years today!" she says, sitting down at the computer where she'll type up his file. She checks in on how often he wears the device, asks about medications. He says he's stopped using his eye ointment. Just doesn't need it any more. "Getting better with time. It's strange," he says.

"Oh, that's great news. Your eye is getting increasingly adapted to it," Duncan says. She murmurs as she types: "Wears device at all times except when sleeping."

"OK," she says, "Let's take a look at your eyes." She turns on a slit lamp ophthalmoscope, which shines a beam of white light directly into his eye.

"Of course, I see that bright light," Lloyd says. "It's not a flash. It feels like a constant light to me."

"And you see it in both eyes or just one?"

"No, I see it in the right eye only."

Duncan moves in to do a closer examination, this time attaching an ophthalmoscope to her head and holding up a larger lens to peer more closely into the eye. "I'm looking at his retina and examining his eyes for health, to make sure that the implant looks healthy and that there's no evidence of any inflammation or problems in the eyes," she says as she works. "Look up high. Look to your left. Thank you. Look down at your toes," Duncan murmurs as she leans in. "Over to your right. Wonderful."

Lloyd quietly does as he is asked, moving his eyes here or there, seeing what there is to see. And after several more minutes of close-up consideration, Duncan renders her verdict on the ocular health of the Model T cyborg and his bionic eye: The only thing worth noting is that he's got an eyelash stuck in it.

FOUR

....

Hearing

THIS CHAPTER HAPPENS entirely inside your head.

Well, at the moment, it is happening inside Aaron Friedman's head. Friedman is lying inside an enormous fMRI scanner in the bowels of a UC Berkeley research building. He's face up, hands tucked to his sides, on the sliding bed that feeds out of the scanner's narrow tube, a visor-like head coil over his eyes. He's propped into position by foam wedges strategically packed under his knees and around the crown of his head, and he's nestled underneath a soft, pellet-filled duvet. All are meant help him lie comfortably—and totally still—while the scanner does its work. Plugged into his ears are huge plastic earbuds, enough to overcome the magnet's electric sawlike drone as it reads out his brain's activity.

As Friedman lies there, he's listening to a podcast, an old episode of *The Moth Radio Hour*, in which writers and comedians share firsthand tales of adventure or woe. Jenifer Hixson is winding up "Where There's Smoke," a story of two women commiserating over cigarettes. In the observation room next door, graduate students Alex Huth and Wendy de Heer notice that the story is almost over and scoot their swivel chairs to their desks. "Are you doing OK?" Huth asks into the mic that connects him to Friedman in the tube.

"I'm fine," comes his voice, somewhat muffled thanks to all the stuff wrapped around his head.

"So we're going to go ahead and play another story, all right?" asks Huth, as de Heer cues up the next track. And with the audio rolling again, they get back to the point of today's study: listening in on Friedman's brain as it, in turn, listens to the human voice. "Using that data," says Huth, "we're going to look at the whole auditory language processing pathway."

Huth is interested in learning about what's going on at the very top of that pathway, where the brain processes semantic meaning. De Heer is more interested in learning how it processes lower-level structures like phonemes, the sounds that make up words, and articulations, or how the mouth creates them. Together, they are working on the bookend process to "writing in," the technology that makes Dean Lloyd's retinal implant possible. They are "reading out" the brain's activity. They use two processes called "encoding" and "decoding," and decoding has some really mind-bending potential applications. If you could become fluent in the brain's electrical language, you could know what anyone had seen or heard. You could also, more roughly, translate that activity *back to the original stimuli.* In theory, this could make possible a new wave of neuroprosthetics that do the opposite of retinal or cochlear implants: turn brain impulses into real-world signals. In the case of hearing, it could make the little voice in your head speak aloud.

Several labs at UC Berkeley—and other institutions—are working on capturing such an accurate model of auditory processing that it might one day be possible to read out internal speech, either displayed as words on a screen or spoken aloud by a device. This would be a brain-machine interface, a device directly controlled by thought. These interfaces are being explored on the motor side of neuroscience as a way to drive prosthetic limbs. For example: You simply think about grasping your coffee cup, and your robotic hand moves.

But they have potential applications for sensory science, too. With hearing, researchers imagine they could aid people incapacitated by strokes, Lou Gehrig's disease, or other paralyses that leave people unable to talk but capable of imagining speech. Others envision them as universal translators: You think the words, the device interprets them. Or, says de Heer, what about controlling your computer or phone, thinking your commands instead of typing or speaking them? "I can easily imagine communicating with devices," she says. "Like a Siri, but you don't have to say it out loud." Or, wonders Huth, what about applications for music? "You just think a melody and then it comes out," he says. "That would be fantastic."

To be clear, so far there is no such device. But there is a cutting-edge technology called stimulus reconstruction that may well lead to one, and it's one of the reasons Friedman is lying in the scanner listening to podcasts. In stimulus reconstruction, the goal is to accurately re-create what a person has seen or heard—without first knowing what it was. Psychologist Jack Gallant's lab at UC Berkeley wowed (and in some cases, terrified) the world beginning in 2008 when they started rolling out experiments showing that stimulus reconstruction could be used in vision to fairly accurately match photos or re-create movie clips that subjects had seen. (Huth is a doctoral candidate in Gallant's lab; de Heer is with a partner lab at Berkeley.)

Gallant will not mind if you call his work "brain reading," because, as he points out, "we are in fact decoding brain activity." They're not reading exactly what each neuron is doing at each point in time—fMRI is too slow for that, and can't read at the level of individual neurons. But from studying people in the magnet, they gain a lower-resolution picture of overall neural activity, allowing them to correlate sensory input with the brain's reaction to it. Then they can build a model showing how one relates to the other.

This kind of modeling, Gallant says, will be a powerful medical tool in the post-Boomer era, as the functionality of the aging brain becomes a predominant health issue. "As we live longer and longer, this is going to be an increasing problem. People are going to physically be OK, but their brain's health is going to be low. And so the only way to address that issue is to do neuroscience research," Gallant says. That fundamental research produces an engineering by-product, he adds: If you can figure out how the brain works, you can decode its activity.

After their success with vision, Gallant's lab began to tackle the decoding of other brain processes, notably language. Once they have learned enough about how the brain processes auditory signals, they think it will be possible to interpret not only what you have heard through your ears, but the imagined speech heard through your *mind's* ear. "I tend to tell people that the most useful brain decoder you could build would be a 'thinking hat' that would allow you to decode the little woman in your head who talks to you all the time," says Gallant. "The minute you can decode that, then it will be a billion-dollar industry, because everyone will want one. You can decode *everything*, right? It replaces basically all other brain-machine interfaces."

But before you can re-create what a person has heard aloud or imagined internally, you need a model of how the brain processes acoustic sound in the first place. Huth and de Heer are interested in speech, so that's why the guy in the scanner is hearing talky podcasts, not, say, a jazz combo or a thunderstorm.

In front of Huth are two flat-screen monitors. Up pop images of Friedman's brain, taken just a few seconds ago. The fMRI scanner tracks the brain's activity by measuring the change in its blood oxygen concentration, volume, and flow. When neurons work, they consume sugar and oxygen, which must be replenished. When fresh blood rushes in to refuel the cells, the magnetic field in the local area changes, and the machine can detect this change. The picture the researchers are getting of Friedman's brain activity is somewhat indirect and time delayed. But unlike using implanted electrodes to record neural signals, this doesn't require surgery, and it lets them watch the activity across his entire cortical sheet, not just a small patch of cells.

Instead of individually measuring the brain's 86 billion neurons, with fMRI, scientists break the brain into an imaginary grid composed of "voxels," or volumetric pixels—all of the neurons in a few cubic millimeters. A voxel represents between a half million and 2 million neurons. Then the researchers measure how active each voxel becomes in response to a certain stimulus.

Today Huth and de Heer are simply recording what's going on in Friedman's auditory cortex. In the slice they're watching on-screen, his gray matter—densely packed with neurons—is rendered as a pale off-white, cut through by darker patches of cerebrospinal fluid and gunmetal clusters of the myelinated axons that join the parts of the brain. The readout displays the change in his blood flow; it's like watching a time-lapse shot of a river delta with the tide washing in and out. "This part right here is busy. It's flickering on and off," says Huth, indicating the dancing lights in the auditory cortex on-screen.

"He's listening," agrees de Heer. "He's doing good."

BEFORE WE GET TOO DEEP into what's happening inside Friedman's brain, let's go over how sounds get there in the first place. The ear's job is to convert sound pressure waves into electrical signals. The outer ear collects and amplifies the waves, then directs them down the ear canal until they hit the eardrum. Then everyone's three favorite bones—the hammer, anvil, and stirrup—further amplify those sounds by acting as tiny levers on one another, concentrating the waves' energy in a small area, a membrane called the oval window.

The conversion of waves to electrical energy happens in a watery environment, inside the cochlea, a snail-shaped structure composed of three fluid-filled canals. When the bones press on the oval window, its movement causes traveling waves to push through the fluid, setting off a series of bulges and membrane displacements throughout the cochlear canals. These at last push a structure called the cochlear partition, and its movements are picked up by hair cells, auditory nerve fibers with tiny dendrites (stereocilia) that extend into the fluid. The movement bends the stereocilia, and the hair cells convert this mechanical energy into an electrical signal, which is passed to the auditory nerve fibers. These signals are then carried to the brain by the auditory nerve.

As with vision, the inbound signal is decomposed along the way. The cochlea starts that process, breaking down complex waveforms into low, medium, and high frequencies, creating an internal representation of the ratio of low to high frequencies that make up each sound. But which breakdowns happen next and at exactly which parts of the brain is still a matter of discovery. After the cochlea, auditory information is passed through seven to ten more synapses—that means seven to ten levels of information processing—before we perceive our experience of the sound.

On each step of a sensory pathway, the neurons there are said to "like" or "care about" or "be tuned for" the kind of information they will pick up and process. Neuroscientists call these qualities the area's "feature space," or what the neurons there represent. In fact, each area probably processes several features, and as the pathway continues, the features become increasingly abstract. With listening to speech, the pathway begins with processing sound by frequency. In the middle, the neurons must care about sound characteristics specific to language, and toward the end, about semantic meaning. But researchers are still trying to figure out where and when each of those things happens, and how the sound is being broken down along the way.

"Not surprisingly, we know a lot about what's happening at the bottom of this thing. We don't know very much about, at each level, what is being done with that information," says Dr. Edward Chang, a neurosurgeon and physiologist at UC San Francisco who studies several aspects of brain mapping, including its applications for hearing and speech. Chang did his postdoctoral work at UC Berkeley in the lab of Robert Knight—like the Gallant Lab, it's a significant player in the world of auditory stimulus reconstruction. Both are part of the Helen Wills Neuroscience Institute, and both are trying to unpack some of the mysteries of auditory processing. Specifically, Chang says of his own interest, "What is perception? That is, what is the distinction between what is coming in through our ears, versus what we actually experience? Those two are often not the same thing."

Just as with vision, the auditory system takes in an overwhelming amount of information, in this case literally noise, from which it must filter meaningful signals. "The system is not just a passive observer," says Chang. "In fact, perception is actually a very active process." And if you can figure out how the brain is manipulating this information, it might clue you in to which features are being processed at each step in the pathway.

For people, a particularly salient type of sound is that special one made only by other people: speech. To help us better comprehend speech, the brain uses a filtering strategy called categorical perception. Chang pulls up a file from an experiment he did in the Knight Lab. On his screen are 14 sound clips, each with a tiny "play" button. He clicks the first one.

"Ba," says a robotic computer voice.

Chang clicks the next button. "Ba," it says again. Three more times: "Ba, ba, ba."

He clicks the sixth button, and now the voice sounds like it is instead saying "Da."

"So it went from ba to da," Chang points out. He presses the next few buttons. Four more times, the computer says "Da."

Then it switches again, this time clearly saying "Ga."

Only it's not that simple. While it sounded like there were only three distinct sounds, each played a few times, there were actually 14. Each was unique, and represented a steady morphing of the "ba" sound into "ga." What our brains should have registered as 14 steps along a gradually changing spectrum was, perceptually, only three. "You have a continuum," says Chang, "but we don't perceive it."

So why not? Your brain wants to sort noise into signal, to compartmentalize sounds into meaningful categories. In this case, Chang thinks, your brain is categorizing sounds based on where they are produced in the mouth. "When you make a ba sound, you close your lips," he says, pressing his together to demonstrate. "When you make a da sound, you use your tongue in the front," he continues, showing how his tongue is pressing against the roof of his mouth. "When you make a ga sound, your tongue goes to the back."

Try it yourself, if nobody's looking. All three feel very different on the tongue and lips. And now, to illustrate Chang's next point, try making something that splits the difference between two of them. You'll probably find yourself awkwardly huffing air. "It's not easy to make something that is kind of in between those, that gives you an in-between sound," Chang says.

Because there is no natural speech sound for something that is in between ba, ga, or da, human languages don't use it. As a result, we don't have percept, or a category, for it in our minds. So when you feed your brain a sound that splits the difference between syllables—possible with speech made by a computer and not a human vocal tract—your brain fudges the fine points, and makes the closest category match. That's a useful trick, because while it sacrifices the fine grain of our auditory experience, it helps us understand what another person is trying to say—and that's what we really care about. (If this idea sounds familiar, it's because it echoes the explanation some taste researchers give for why it's so hard for us to isolate a sixth taste, why people default to existing descriptors like "bitter" when tasting contenders like fat and calcium. Since we don't already have a percept in mind for a new taste, they think, we leap to the nearest category match. We are as insensitive to the differences between, say, bitter and calcium-y as we are to the subtle gradations of a sound somewhere between ba and ga.)

Figuring out the relationship between speech sounds and brain activity is an important part of Chang's work, one that's taken him deep inside the brains of living people. At UC San Francisco, his lab's experiments focus on the superior temporal gyrus, part of the temporal lobe, which is the pinnacle of the auditory cortex hierarchy. "There is something about this area that is very fundamental to the processing of language and speech and this thing that we call perception," Chang says.

While the Gallant Lab uses fMRI to listen in on the brain, Chang uses surface electrode electrocorticography, or ECOG, in which an array of electrodes is placed directly on a person's brain. In addition to studying audition, Chang works with epileptics, and this gives him access to a very special clinical subject pool. Some patients must undergo a craniotomy, in which their skull is cut open and an electrode array is placed on the brain. The array lies on its surface, but doesn't penetrate it; an array can be anywhere from about the size of a quarter to ten centimeters square. The patients wear their array for about a week as they wait in the hospital for a seizure that the electrodes can record, mapping its point of origin. This will help their neurosurgeons know where to operate.

The wait can be long and dull, and at UC San Francisco and other clinics, people often volunteer their down time, gamely participating in all sorts of cognitive tasks—reading, talking, moving around, solving puzzles—while researchers take advantage of this rare opportunity to record the human brain in action. While much more invasive than fMRI, ECOG has an advantage: Instead of indirectly sussing the brain's activity, it can read, in sharp detail and real time, what large communities of neurons are doing when exposed to a stimulus. That timing is particularly important for studying speech, because speech sounds change rapidly.

Chang pulls up an illustration taken from a study in which subjects wearing arrays listened to spoken phrases. Chang and his colleagues painstakingly tried to figure out, electrode by electrode, what the neural pattern of activity was, or which sound characteristics these neurons in the superior temporal gyrus "liked." They concluded that the neurons were activating in response to the sounds' articulatory features, or how people move their mouths to make them. "The auditory system for speech seems to be very tuned to the sounds that are actually generated from the vocal tract," says Chang. "What are the things that are going on with your lips, your tongue, your jaw, your larynx?"

He points to a list of the sounds that the neurons at a particular electrode seemed to be responding to: the familiar ba, da, ga. Now while you form these sounds in different ways, they share an important characteristic: To make them, you close the vocal tract, blocking airflow. People who study speech refer to these sounds as "plosives."

Chang points to the results from another electrode: This one was most sensitive to "z," "f," and "sh" sounds, or fricatives—sounds created by forcing air through parts of the mouth.

And so on down the list—neurons in different areas were paying attention to sounds produced in different ways within the vocal tract. That's significant, he says, because, more than alphabetic letters or even consonant and vowel sounds, these are the building blocks of speech. "These cues are the ones that define every language in the world," he says—they're like a periodic table of sounds that can be combined to create an infinite number of meanings. So if his lab is correct, that's one blank spot on the auditory pathway map filled in, one level where we know how our neurons are filtering the sound world for us.

BRIAN PASLEY IS INTERESTED IN FILLING in spots on the map, too. Pasley is a soft-spoken postdoc in the Knight Lab at UC Berkeley, where Chang also researched, and like him, spends a lot of time trying to figure out what the brain is keying into at each stage of the pathway. But he's also interested in potential engineering applications, and in 2012 was the lead author on a collaboration with Chang and others that essentially broke the idea of internal speech reconstruction to the public.

This paper was actually supposed to be a fairly wonky one, a comparison of two different decoding strategies. But it turned a lot of heads, partly because the group had reconstructed a reasonably clear version of what their subjects had heard (and posted sound files on the Internet to prove it), and partly because their paper ventured some ideas about how the technology might lead to next-generation neuroprosthetics. Pasley envisions a device that would help people who have, say, locked-in syndrome or Lou Gehrig's disease. "Someone who is paralyzed, has no ability to speak—well, in most cases their language system is still functioning just fine," he says. His lab's initial decoders were developed for people *listening* to speech, but in 2013 they began testing decoders on *imagined* speech. And that makes him think you could ultimately build an inner voice reader for people who cannot talk. "Basically, they would be imagining the message they want to communicate, and we could use a similar approach to reconstruct their intended speech," he says.

Pasley's work is built on ECOG readings, as well; he gets his data from patient volunteers at UC San Francisco and other institutions. For the 2012 study, Pasley used data from people who had the arrays laid over their superior temporal gyrus. A good deal of his work has focused on how this area is processing frequency or, essentially, pitch. Think of this space as a 3-D piano keyboard, where neurons in different locations are tuned for higher or lower frequencies. When we listen to people talk, the frequency content of those sounds changes over time. As it does, neurons in the different areas, sensitive to specific frequencies, either activate or don't. That produces a brain activity pattern. Now imagine that the neural keyboard is really more of a player pi-ano: If you knew the pattern in which the keys were being depressed, you could eventually call the tune.

This was the theory behind Pasley's study. His volunteers had a simple task: As they sat in their hospital beds, they listened to prerecorded words played through speakers. Their electrodes relayed how each part of the superior tem-poral gyrus responded to each sound. That created two neatly matching data sets. "We know *exactly* what the patient was listening to. We know *exactly* what the brain signals were," Pasley says. "So given a particular sound, we have measured what that sound causes the brain to do, in terms of its electrical activity."

"So we have input and output," he continues. "And we want to build a sta-tistical model that relates those two." First, you create an encoder that goes from stimulus to brain response. (Math people: It is essentially a linear re-gression that assigns weights to the various stimuli, then multiplies them to optimally predict the evoked brain activity.) If you have been paying very care-ful attention, you will recognize that this kind of stimulus-encoding process is how Sheila Nirenberg's lab mathematically converts images into ganglion cell code, the basis for their vision prosthetic system.

Then if you want to see how well the encoding model works, you try it in reverse. You decode, or use the neural activity to reconstruct the stimulus. In this case, Pasley checked to see if his model could re-create words that, when spoken aloud, sound right to the human ear. (Math people: You test your encoder by checking how well the response you measured at each voxel or electrode matched the original input. Then, to create the decoder, you do another linear regression, summing across the voxels and weighting each response by its likelihood, or how well it predicts the original stimulus.)

Pasley taps a few keys on his computer, and brings up his lab's first efforts at speech reconstruction. Each word plays three times—the first is the origi-nal the patients heard; the next two are reconstructions using the two decod-ers they were comparing.

"Waldo," says a woman's voice clearly.

"Waldo," says what appears to be an underwater robot.

"Waldo," says what appears to be an underwater robot gargling sand.

The same thing happens with three more words: "structure," "doubt," "property." The reconstructed words are intelligible, but garbled and metal-lic. "Honestly, you have to really hear the real sound first," Pasley says with a laugh. "That helps a lot. If you just heard that without a reference point, it's probably not recognizable. But if you hear the real thing first, you kind of know what to expect and you can identify some coarse similarities."

Now that seems modest, but the big picture here is the proof of concept, not clarity of sound. That's essentially a math problem, one that will improve as researchers build better statistical models. And as Pasley points out, they're not even sure if the superior temporal gyrus is the best place to figure out how the brain is sorting sound frequencies. That process may have begun earlier in the auditory pathway, deeper in the brain's interior, but it's just too hard to squeeze electrodes in there.

So far, Pasley has shown that you can record the neural activity in some parts of the auditory cortex as people listen to sounds, build a model that correlates the two, and re-create (warbly versions of) those sounds. But that doesn't quite get us to Gallant's idea about reading out internal speech, what he calls the "little woman in your head," or what Pasley refers to as auditory "imagery." And that's a big leap, from acoustic sounds to imagined ones.

"Imagery is tough to define," Pasley says. "By definition, it's an internal subjective experience, right?" But it's essentially what you are doing right now as you read silently to yourself, converting the text on the page to an internal auditory representation. Or it's that little voice you hear when you give yourself a pep talk in the mirror or wonder if you left the oven on. "Just like visual imagery, where we can imagine faces or a scene, I think most people have some level of auditory imagery," Pasley says. "We can imagine the tune of our favorite song or the most annoying jingle—sometimes involuntarily."

The technical terms for these are "overt" speech (what you say with your mouth) and "covert" speech (the voice you hear in your head). In 2013, Pasley's group began a new series of experiments exploring the reconstruction of covert speech, this time asking the subjects to read some text aloud, and then read it silently, trying to produce that internal auditory imagery. He brings up a few examples of phrases the people read: "That makes me happy" and "I am hungry." Pasley's group then began working on a decoder model based on these volunteers' brain activity from when they read aloud. But then the twist, Pasley says, is that "instead of applying it to data we record when they are *hearing* stuff, we apply it to data we record when they are *imagining* stuff."

To check whether this works, the researchers tried to match the reconstructions to the original sentences. This kind of task, called "identification," simply tests whether you can make a match within a limited pool of choices. Think of a magician who asks you to pick a card from his deck. No matter which you choose, the correct answer must be one of those 52. "The basic idea is to prove to ourselves that what we're reconstructing is not just total noise, meaningless patterns," says Pasley. "We want to be able to use those patterns to actually pick which sentence was being imagined. And at least with this first bit of data, we can do that."

They reconstructed the sounds, too. Pasley presses the button. It plays the phrase "This makes me happy"—first the overt reconstruction, then the covert one. Then he plays back the "I am hungry" version. The overt reconstructions sound like our sand-gargling robot pals, distorted but intelligible. The covert ones are much harder to make out. The syllables jam together, making individual words into long sounds that often distort into an electronic bleat. But there's something distinctly speech-like about the sentences; they rise and fall with the intonation of the original; here and there a vowel or consonant sound clearly stands out; the overall "shape" of the phrase just feels right. Still, there's no way a listener coming in cold would understand these lines. But at this point in the data analysis—just a few more patients to go—Pasley thinks they can make matches at a rate better than chance. "We think there's something there," he says. (By the next spring, his group will have concluded that they could, on average, for the seven patients who participated, although the rate was only slightly greater than chance.)

Now this might seem a bit speculative or ethereal, given the primitiveness of the audio playback quality. But it's also something that has already been done, many times, and in the next building over on campus—just with a different sense. To get an idea of where Pasley is headed with hearing, we have to go back to the Gallant Lab, the one that is putting people in the scanner and feeding them podcasts. They started dazzling the world with their stimulus reconstruction projects several years earlier—but in vision.

DR. JACK GALLANT IS A WIRY FAST-TALKER who easily whips comic one-liners into what is otherwise a steady stream of high-level neuroscience. When we meet in his group's lab space, he's wearing a cream pullover sweater, jeans, and blue tennis shoes, his dark brown hair worn slightly shaggy. He doesn't shy away from the difficulty of what his lab is trying to accomplish, given how little is known about how the brain's neurons are organized to handle sensory input in the first place. "We are essentially trying to reverse engineer the brain. We don't necessarily know what the parts are," Gallant says. "Think of it this way: Someone has given you this meat, and there are a bunch of areas in this meat. How the hell do you discriminate between them?"

In vision, where Gallant spent most of his career, this is a particularly thorny problem, because the vision system has a ton of meat. Primates have between 32 and 40 visual areas, and humans likely have more, thanks to their larger brains and capacity for language. These are spread out over the occipital cortex, and the occipito-parietal and occipito-temporal areas. The first six of these broad areas are known as V1 through V6, but there are dozens of others.

As with audition, each of these areas likely represents multiple features. Just in V1, or primary visual cortex, the neurons process 10 to 15 features, including the image's X and Y position on the retina, its eye of origin, its orientation, object size, and temporal frequency, or the rate at which things change. As information moves along the visual pathway, which visual features are represented is less well understood. Altogether, Gallant thinks the brain represents well over 1,000 visual features.

Now, he continues, you might ask why there are so many visual areas—or areas for processing any sensory signal, really. "There are multiple possibilities for this. The first is, well, neurons are really stupid," he deadpans. A neuron, he says, is basically a simple analog-to-digital converter, with a dendritic arbor on one end where it takes in information. Then it makes a decision and fires an action potential down along its axon, passing the message to the next neuron. Maybe if you want to do a complicated computation, you need a really long series of them. Or maybe it's just that evolution built the brain inefficiently, with all sorts of redundancies and add-ons. "So if you are a really dumb animal and you only have a V1, now you want to do something more complicated with vision—well, what do you do?" he asks. "Do you change V1, or do you just spurt off a new visual area and send it to there? And that's probably what happened."

It's no surprise, Gallant says, that we don't yet know the ins and outs of such a complex system, built randomly over time and without marked parts. But you can get an idea of what the parts *are* by figuring out what they *do*. "Our job as neuroscientists is to try to discover what the feature space is, given the pixels that we have put in and the activity that we measure," he says. In other words, you show someone pictures, you see how their neurons respond, and you guess what they saw. And then if you have a good enough model for what they saw, you can try to re-create it.

By now, that should sound a lot like stimulus reconstruction.

To be very clear, not all brain reading is stimulus reconstruction. Gallant breaks it into three levels. The simplest is classification: You show someone an image, and based on their brain activity, guess if it fits into some broad category. "Imagine I give you a deck of cards, and I tell you, 'Take a card out of the deck and go into the magnet and look at the card and then come back out,'" Gallant says. "I say, 'OK, I'm going to guess whether it's a face card or not a face card.' That's classification."

The next level is identification, in which you guess given a limited set of choices, like the magician picking your card from a deck of 52. "Identification would say, 'OK, I'm going to tell you whether it was the jack of diamonds or the ten of clubs,'" Gallant says. That's exactly what Pasley was doing by trying to match reconstructed covert speech to the original sentences. And that's where Gallant's lab started, too, using still images. In 2008, they showed they could reliably do identification matches with subjects who had seen grayscale photos of real-world objects like food, animals, and outdoor landscapes.

Then they moved on to stimulus reconstruction, the most difficult level. This is much harder, because the original input can be anything—and you don't get any clues beforehand. "In reconstruction, I would say, 'Well, I don't even know if you looked at a deck of cards. Maybe you looked at a bunch of photographs. I have no idea what you actually looked at," Gallant says. That's what Pasley was doing with the experiment that led to the sand-gargling robot versions of the word "Waldo," and that's what Gallant's lab has been doing almost eerily well over the last several years.

Gallant pulls up a 2009 study, in which subjects looked at photographs, and the lab tried to reconstruct what they saw. The tricky part about reconstruction is you don't get a preexisting set of images to use when you try to make your match. Instead, you need a second body of photos, totally different ones that don't overlap with the first set. (This second set is called the "prior.") The bigger and more random the prior, the better—it gives you a better chance of finding an image that is really, really, really *like* the original, without being exactly the same photo. So for this study, the lab built a model that would pick the closest approximation from a prior of 50 million photos.

Gallant shows a few of this study's best matches: A harbor is matched with a similarly shaped bay; a line of theater performers is matched by what looks like a group of kids lined up on steps. Neither would be mistaken for the original photo, but they're in the right ballpark.

But that wasn't always the case. In the first phase of the experiment, the model used only brain activity related to a fairly low-level visual feature: its spatial qualities, which included contrast, orientation, and spatial frequency. Gallant pulls up a match made in this round. The original photo is of two buildings. The match is of a dog. Now, that doesn't seem right at all. Except that if you peer closer and look only at the *shape* of the objects, and forget about what they *mean*, it starts to make sense. The open space between the two buildings is shaped like the dark area above the dog's head. The dog is sitting in front of a bedspread with a square pattern that mimics the placement of one of the building's windows. Some of the edges of objects in the two photos correlate pretty well. As an extremely literal interpretation of shape, the match kind of works. But in terms of significance, it doesn't.

So in the second phase of the experiment, they also matched using a higher-level feature, the semantic class of the object—in other words, was it an animal, a vegetable, a building? Gallant pulls up two photos the algorithm matched using only their spatial qualities: a bunch of grapes and a photo of an adult hand touching a baby's fingers. Again, not a great match. Then he shows how they were rematched by the model when the semantic information was added. Now the grapes are matched to a photo of some mushrooms, both clusters of round edible objects about the same size. That's not right, but it's better. The accuracy of the match has been improved by correlating just two kinds of features: spatial and semantic. If Gallant's estimate is right, there are hundreds more features you could tune to further improve the picture.

With these photo studies, they were getting better matches, but Gallant wasn't quite satisfied. "Who cares how the brain responds to still images?" he asks. "That's not natural." So in 2011, the lab switched from feeding their subjects photos to showing them videos. Dr. Shinji Nishimoto, then a postdoc, now at Japan's National Institute of Information and Communications Technology, decided the subjects should watch movie trailers. He reasoned that they'd do a good job of mimicking the movement we see in daily life, and be less boring than the experimental stimuli often used in fMRI research—geometric patterns or parts of faces. "Watching movies in the scanner was relatively fun," Nishimoto emails from Japan, "compared with watching a flickering checkerboard for many hours." (Researchers often become their own lab subjects for fMRI projects, because it's hard to find volunteers who will lie in the scanner for so long.)

So Nishimoto and two other lab members watched the trailers, and built a model based on their brain activity. Then it was time to try reconstructing. To assemble their prior, they set up a computer program that did nothing but download from YouTube until they had 5,000 hours of random videos. ("You might ask: Why 5,000 hours?" Gallant says. "Five thousand hours is about the amount of hours you are awake during one year. So you can think of our prior as one year of your visual experience if all you did during that year is watch YouTube videos." Imagine that and try not to weep.)

Then they ran their model and tried to reconstruct the originals using this huge video array. Gallant cues up a side-by-side comparison of clips from the movie trailers and their nearest video matches. "You can see that sometimes this thing works well and sometimes it works poorly," says Gallant. On the left is a video clip of a bleeding ink spot, and on the right is its unconvincing match—a multicolored blur. "Here it is working horribly, and the reason it is working horribly is there is nothing in our prior that looks even remotely like that random ink spot," Gallant says.

Another pair comes up. "There is nothing in our prior that looks like this elephant," Gallant says, and indeed, the model has matched the elephant with a rooster.

Then it matches a bird with Eddie Izzard.

And then it starts doing better. It's good at matching people to people, getting their general shape and position right. It matches a video of Steve Martin as Inspector Clouseau with a clip of Adam Savage, the host of *Mythbusters*. They are both men, both standing, and they both appear on the same side of the screen. It's also good at matching up close-up faces with similar ones, or text with text. "Right," says Gallant, "because there is a lot of text on YouTube. There are a lot of faces. If we had a cat, there's craploads of cats on YouTube and it could find cats. So if this thing is really common in our prior, we can find it really well. If it's rare, we can't find it." And, he points out, with a bigger prior they would be more accurate. "Our prior is only 5,000 hours of video. If we had 50 million hours of video we would do much better."

Gallant says he was pretty satisfied with this as a proof of concept, but Nishimoto wasn't. So Nishimoto developed something he called the "average high posterior," a way of taking the 100 best-matching clips and averaging them to see if together they would form a better match than a single image. Each clip is different in some ways from the original, Nishimoto writes, so "by averaging the top 100 clips I tried smoothing out these deviations. I don't think that our current decoding results were really accurate, but I think that these were not too bad as the very first step."

Nishimoto might be being modest. As Gallant rolls back some of the reconstructions made this way, it's pretty extraordinary. He plays an original clip of elephants walking along the desert sand. The reconstructed image playing alongside it is like what you might get if you were watching that image through a smear of Vaseline. You can't tell that the elephant is an elephant, but you can tell that it's a living creature of a certain size and rough shape, that it's moving from left to right at walking speed, that there is a sky in the background. And it's right around here that you will, depending on your engineering and philosophical bent, start to get either very excited or very frightened, because the reconstruction is starting to look pretty dang like the thing that is being reconstructed.

There was another idea the lab wanted to try, and this became their bridge into the work they are now doing with hearing. Led by Huth, who is interested in semantic function, they took the same collection of movie trailers, and had five subjects watch them in the scanner. This time, they tallied up how often 1,705 common objects and actions appeared in the trailers. Then they built a model that correlated the brain's activity with the appearance of those objects and actions. From that, they developed a model of what they called the brain's "continuous semantic space," showing that categories the brain represents similarly (like "animal" and "dog") are close together in the model, while categories it represents very differently are far apart.

Gallant calls up those side-by-side clips again. On the left are the original trailers. This time, on the right is a word cloud, in which words related to the images in the trailers bubble up and fade away depending on what's happening on-screen. So to the left, we get a few seconds of a rom-com starring Anne Hathaway, in which she seems to be chatting with a group of friends. To the right, "we get 'woman,' 'man,' 'talk,' 'room,' 'walking,' 'face,'" Gallant narrates as the words pop onto the screen—all pretty good descriptors of what is going on. The video clip switches to something filmed underwater. "Now the ocean comes, so we get 'fish,' 'swim,' 'water,' 'ocean floor,' 'body of water.' This is a manatee," he says, pointing at a blimp-like animal swimming by. "We don't have that, but it gets 'whale,' which is pretty close."

Then the model gets stumped a few times. It mistakes an expanse of silvery snow for water. It defaults to "building" for a tanklike Arctic snow crawler. As we watch Kevin James as Paul Blart, mall cop, pratfall into a set of glass doors, the model picks up "room," "walk," and "building," all correct, but misses with "road," which is how it misinterprets the long mall corridor behind him.

Once the lab had started working with modeling how the brain processes semantics, it wasn't a huge jump to studying speech and hearing. Stimulus reconstruction, Gallant says, works on any kind of stimulus. "We used vision as a platform for developing the technology, and now it's basically done. So now we can apply it to everything," says Gallant. "We can apply it to audition, to language. We can apply it to decision making, memory systems—all of the other systems of the brain can be essentially attacked using the same kind of very sensitive technology."

So then, I ask, is the endgame to be able to reconstruct *anything* the brain does? "Well, I'm a megalomaniac," Gallant begins drily, only to be interrupted as loud laughter erupts from a grad student hidden in the next cubicle. "I want to predict all of the brain activity in any situation that you are ever in. That's the goal," Gallant continues, undeterred.

"You might think he's kidding!" the grad student shouts invisibly from behind the cubicle.

I do not think he is kidding.

"Yeah," says Gallant wryly. "He's not kidding."

THE INCREASING ACCURACY OF RECONSTRUCTION opens up some really big questions about how far this technology can actually go, and how it might be used. And as Aaron Friedman, still in the scanner, listens to his final round of podcasts, Huth and de Heer are in the next room debating one of the field's big unknowns: whether your brain activates for imagined stimuli as it does for real stimuli. In other words, does it listen to the mind's ear as it does your physical ears?

The first generation of reconstruction models were based on subjects sensing actual sounds and images, not imagined ones. Similarly, models that are based on having subjects speak aloud may not translate well to what happens when they speak silently in their heads, because out-loud speech involves motor activity. Your tongue and jaw must move, and that generates its own neural chatter. "Imagined speech and produced speech might be really different," says Huth. "But that's something that's really hard to get at with fMRI, because speaking causes terrible, terrible problems for us." Moving your jaw changes the magnetic field around your head and jostles your brain, making the image harder to read.

Speech also relies on fine timescales. "The order in which words are said makes a huge difference to what they mean," says Huth. "Or," adds de Heer, "the order of phonemes within the word makes a big difference to the word." And fMRI, at least, is not a great tool for capturing such rapid changes.

On a practical level, Gallant says, the neural activity for imagery and for sensory perception can't be identical. "The way your brain works when you do imagery *must* be different than the way the brain works when you just look around the world," Gallant says, because if those were exactly the same, "then you wouldn't be able to tell the difference between an image and the world and your ancestors would have been eaten by a tiger." We would exist in a sort of perpetual daydream, caught between incoming sensory stimuli and the figments of our imaginations. The same issue applies to internal speech, he continues: "Your brain can't be doing the same thing as when you actually hear voices, because if it did, you would think that the voices in your head were coming from the outside world." So our brains likely have mechanisms to tell which is which. Some research in primates indicates that auditory cortex activity is suppressed while the animal is vocalizing, perhaps a way of signifying that the sound is coming from the self.

But these likely differences haven't stopped anyone from trying to reconstruct the imaginary, and it will probably not surprise you that Gallant's lab has already tried something like this with vision. In 2014, led by lab alum Thomas Naselaris, now at the Medical University of South Carolina, they tried reconstructing remembered images. They asked each subject to memorize five works of art, including the *Mona Lisa*, staring at them repeatedly to make perfect mental images. Once inside the scanner, the subjects were just cued with words, and had to recall the images. Then the lab tried to reconstruct the images based on the subject's brain activity as they recalled the portraits. "And the question is," Gallant asks, "How well can we do that?" He pages through some results: The decoder matched the brain activity from recalling the *Mona Lisa* to a photo of Salma Hayek. It matched a cat to a dog. It matched some vegetables to other vegetables. Overall, says Gallant, it worked about a third as well as decoding images people were actually *seeing*, not remembering.

Yet reading out the imaginary could be an incredibly powerful tool, and not only for the medical applications Pasley envisions, like helping people communicate. Imagine being able to draw despite your lack of fine motor control, or to compose without having perfect pitch or a pleasant singing voice. You just think, and the image or sound is rendered by computer. (Gallant calls this "brain-aided art.") Or, as they suggest in the *Mona Lisa* study, you could use image recall as a way to search the Internet, looking for remembered photos instead of keywords. Really, you could drive almost any device with your mind. "It is ridiculously awesome," says Huth. "This is like sort of a science fiction dream, to be able to just communicate your thoughts directly into a computer."

But reading out is also potentially scary; you can just as easily imagine a *Minority Report*–esque era of "Precrime" and thought monitoring, when what happens within the confines of the skull is no longer private. What if others could scan your head and read out information you don't want to reveal? Do any of us really want a robot box announcing our every thought?

Yet there are two big hurdles reconstruction technology would have to jump before such sci-fi scenarios become reasonable. The first has to do with abstraction. So far reconstruction does OK with re-creating actual sensory input, and less well with rigorously memorized input. But many thought processes are much more vaporous than either of these, and therefore harder to capture.

Let's start with dreams. So far, it seems possible to identify some elements of dreaming, a time when the brain's visual areas are highly active. In 2013, Yukiyasu Kamitani's lab at Kyoto's Advanced Telecommunications Research Institute International announced that they'd been able to identify the classes of objects three subjects had seen during hypnagogic imagery, the dream-state that precedes deeper REM sleep. (Their model correlated fMRI activity between when the men were awake and watching videos, and when they were asleep in the scanner. They also kept waking the men up to ask what they were dreaming, since dream recall is spectacularly wispy.) This was a classification task—they were able to make categorical matches, but could not reconstruct the action or imagery of the dream itself.

Reconstructing ordinary memories—not images you have repeatedly stared at—presents a bigger problem. Your mental image is only as accurate as your original memory, and that's not necessarily great. Gallant is often asked whether reconstruction could be used to tap into the memories of crime witnesses. His answer is a resounding *no*. "Eyewitness testimony is notoriously unreliable," he says. Your memory isn't Kodachrome, faithfully catching the light. You're recalling your *perception* of what you saw or heard, and that is subject to bias, error, emotional warping, and fading over time. "If you just do brain decoding, you're just going to get whatever lousy answer the person would have given you anyway" if you'd asked them on the witness stand, Gallant says.

And it would be harder still to read "thoughts," whatever those are. Reconstructing internal speech, Pasley points out, "completely depends on you having a vivid auditory component of your imagery." But thoughts are often vague intentions, judgments, or desires you never consciously verbalize. Your internal dialogue is a level above that, Pasley says, "the *translation* of the thought." While reconstruction may be able to get at the translation, it's not clear if it can read the original.

The second big hurdle is how much effort it will take to get inside your head. Today, reconstruction cannot be done without your permission. All current technologies are invasive, time consuming, and decidedly nonsneaky. They require a radical surgical procedure or a real commitment to lying still in the magnet. This is some very hard biohacking, and requires a tremendous degree of technical skill and oversight. "This is only useful for medical purposes for someone who has no other option," says Pasley. "It's not as simple as just putting some electrodes on the scalp."

But Gallant likes to refer, tongue in cheek, to a future invention he alternately calls the "iHat" or the "Google Hat," some wearable brain-reading device that is who-knows-how-many decades off on the consumer product horizon. Personally, Gallant has mixed feelings about this prospect. "Our internal thoughts are our most private stuff that we have, and the thought that you could basically build a decoder to decode internal thoughts, I find to be really, really exciting and really, really scary at the same time," he says. It's easy to imagine abuses of power. For example, he says, "imagine that the police have something that looks like a radar gun, and they point it at you and it decodes your internal speech. That's not an impossible thing to think about, because your brain is just a computer that's giving off signals." Sure, he adds, there's no technology today that can read the brain at a distance, but that doesn't mean it can't happen. Before we get anywhere near a Google Hat, he says, "there are a lot of really serious, very deep ethical issues that are going to have to get dealt with."

How would you control who reads out your information, and how it would be used? We'll bore deeper into questions about privacy and surveillance in Chapters 10 and 11, after we've taken a look at other technologies that link perception and machine. For now, it's worth pointing out that we live in an era not only of increasing government encroachment on privacy, but of the voluntary sharing of personal information, including biometric data gathered from devices like smartwatches, wristbands, and cell phones. But we also live in a time when many people could be helped by prosthetics that read out brain activity. Every year in the United States, about 800,000 people have strokes. Another 12,500 suffer spinal cord injuries. Some 5,600 are diagnosed with Lou Gehrig's disease.

Learning to read out and write in are still new technologies, but they are headed toward the same point: being able to do them simultaneously, to create a fluid loop between person and computer. That's what would make brain-machine interfaces helpful in ordinary life, and is perhaps the strongest argument in their favor. And that's the research turf we're headed to next—it's a story that ends with people, but starts with robots.

FIVE

·································

Touch

DR. SHERRY WREN IS MIDOPERATION, everything but her brown eyes hidden behind a sterile blue surgical gown, hairnet, and face shield. The operating room is dark and cool, lit mostly by the glow coming from video monitors displaying a camera feed from inside the patient's body, so the surgical team can watch the tips of Wren's instruments as she works.

"You can actually see the gallbladder now," says Wren, indicating a pale, slug-shaped organ. It's got to come out, because it's laced with stones. The smoother purple structure above it, that's the liver, which Wren doesn't want to disturb. The yellow fluff surrounding everything is fat, which Wren begins to efficiently tease away from the gallbladder. Once she's disentangled the organ, she'll clip off the duct that connects it to the body, and it will be ready to go.

Wren deftly works a grasper with her left hand, and with her right, an electrocautery hook, which separates tissue by electrically burning through it. A second surgeon, Dr. Arghavan Salles, is standing at the bedside by the patient's hip, holding an additional grasping instrument to retract organs so that Wren can work around them. But Wren herself is actually on the other side of the room. She's interacting with her patient through the arms of a robot, which curl gently over the patient's body as she lies sleeping on the bed, everything but a small square of her stomach veiled by blue paper drapes.

The robot is more formally known as a da Vinci Surgical System, and it's one of the first commercial robotic surgical tools in the world. It's made by Silicon Valley company Intuitive Surgical, and since it launched in 1999, it's been increasingly used to help surgeons do minimally invasive procedures on tissues deep in the body—the pancreas, heart, intestines, reproductive organs. With the robot, Wren can enter the body in a much more contained, precise way than she can with her own hands. As she puts it, "It works great in a small, dark hole."

The surgeons currently have three instruments, plus a camera, inside the patient's body, all going through a 2.5-centimeter incision in the navel. This will result in what Wren calls a "beautiful scar," one that will "look just like a belly button." These instruments feed into the body through a port, an hourglass-shaped piece of silicone that holds the incision open. Four trocars, or long tubes, have been squeezed through the port, and the instruments are introduced and withdrawn through them. The instruments have flexible stems that lock to the robot arm that drives them, and terminate inside the patient's body with the graspers and hooks, which Wren manipulates from her master console in the far corner of the room. The patient's belly has been puffed full of carbon dioxide, swelling it to almost beach ball size. Between the gas, and the triangulation of the instruments, and the light on the camera, the surgeons now have an open, well-lit area in which to work, despite the tiny incision. The robot arms look like the appendages of some kind of cyberspider, and as Wren steers them, they dance serenely above the patient, a light on each arm blinking white or blue in the darkened room.

"Alex, can you get a clip applier ready?" Wren calls across the room to Alexander Lao, the scrub nurse. If Wren had been standing at the bedside, he would have handed her this instrument, a tiny set of jaws holding a white plastic snap. When Wren closes its jaws over the duct, they will clamp together, and the clip will detach, sealing off the duct. Instead, Lao attaches the clip applier to a robot arm, and sends it down a trocar tube. From Wren's point of view at her console, one of her instruments disappears and the new one loads into place.

Wren controls her instruments by gripping a manipulator in each hand, pinching it between her thumb and her middle finger. These are geared so that she can move freely, naturally rotating her wrists as she scoops with the hook and opens and closes her fingers to squeeze with the graspers. As she dips and curls and twists her hands in midair, her movements echo those she might make if she were holding tools by hand.

She doesn't directly see herself doing this. Her console is enormous and arcade-like; she works with her head directly inside and her hands below it. She sees only the video feed of her instrument tips moving. The screen renders to her a brightly lit, 3-D and larger-than-life image of the body cavity in which she is operating. It is a strangely beautiful world, glistening and richly colored in yellows and pinks, shot through with sparks and puffs of white water vapor that billow up when she touches her electrocautery hook to flesh. But as deeply visual as her world is right now, Wren has no sense of touch. And at Stanford University, where Wren is a professor of surgery, researchers are exploring whether—and how—it's possible to give that back to her.

Touch is incredibly complicated, encompassing dimensions like pressure and texture. (Some sensory scientists consider pain a subset of touch, since the nociceptors that react to painful stimuli are embedded in the skin, but as we'll see in Chapter 7, pain is also a multisensory perception.) Touch's many components are difficult to render through a machine interface, especially at the delicate levels required for surgery. This problem is being tackled first in the older field of robotics, but what people discover here will have profound implications for the developing field of human neuroprosthetics. Instead of controlling a robotic arm at a distance and through a console, as Wren is doing now, these robotic limbs will be attached to the body and controlled by thought. One of the questions before the field is whether they can endow robotic limbs not only with fine motor skills, but the touch sensitivity so exemplified by a surgeon's hands.

These two fields have a shared vision: the marriage of writing in and reading out, making them happen seamlessly, so that the motion seems completely natural, and you can feel through the device as if with your own hand. The robotics term for that is "transparency." The neuroprosthetics term might be "low latency." Here, in this surgery, half of that problem has already been solved—the half that translates the surgeon's movements to the machine in real time. Wren can move so smoothly, she says, that "I don't even think about it. I have no idea what my hands are doing."

But to get information back to her, allowing her to sense how the environment of the body has responded to her touch, for now robotics designers have to take a sensory shortcut. Without touch, Wren relies almost entirely on vision. "Do I feel?" she ponders aloud as she works on the on-screen gallbladder. "The funny thing is, is that it's such an immersive visual field that it feels like I feel."

"Your brain is fooled," she continues. "My brain sort of makes me think like I'm feeling something."

But, well, *how*? "I don't know," she says, sounding puzzled. "It's almost like I feel like I'm doing something. I *almost* feel it. It's hard to describe."

THE PHRASE WREN IS SEARCHING for might be "sensory substitution," when one sense fills in for another. It's a concept that Dr. Allison Okamura's lab is considering as they search for ways to convey touch to surgeons—and perhaps give them tactile superpowers. Okamura, a professor of mechanical engineering, runs Stanford's CHARM Lab (Collaborative Haptics and Robotics in Medicine), which has a partnership with Intuitive Surgical. A da Vinci console carcass stands in the middle of her lab, a metal skeleton without screens or plastic casing, a rigging for them to build atop.

Sensory substitution, Okamura says, "is why the da Vinci still works even though it doesn't really have haptic feedback. People just learn to use visual information." But, she adds, it's not always enough. "There are going to be times where you want to do something where visual information just doesn't have what you need. And I think that has limited the proliferation of surgical robotics into certain types of tasks." For example, surgeons using the da Vinci work in a world with no force feedback, so when they press a tissue with their instruments, they cannot feel its resistance. While surgeons learn to read visual cues about how much force they are applying, Okamura says, force feedback does more than just signal pressure. It actually influences the way your body moves. If you lean on a table, you don't fall over, because the table is pushing back on you. If you pick up an object, the force feedback from its weight tells you how hard you need to grip it. "Getting force feedback," she says, "keeps the physics real. It makes sure that people do the right thing." That's important when you don't want to, say, accidentally push too hard through tissue.

Okamura wants to see if she can bring some of that realness back. "How can you make the user feel the sense of telepresence, as if they are not manipulating through this complicated set of linkages, but that it's almost as if they are directly touching tissue?" she asks. That's still a way off, she says—current technology can't re-create what a surgeon can sense by hand. "But eventually haptic technology will get to the point where we can be enhancing [the surgeon's] sense of touch, and not even just trying to get it back to him to the way it was before."

Okamura has a special interest in designing robots that work with people. As a student, she'd been attracted to mechanical engineering because she liked working at the "human scale," rather than the atomic and cosmic scales of physics. "What could be more human scale than things where a human physically interacts with a robot?" she asks. In grad school, she worked on enabling robot fingers to touch, but the sensation wasn't conveyed back to a human operator. Then she got a part-time job with a haptics company that was developing touch feedback for medical training simulators. One of her projects was a mannequin head for practicing sinus surgery; the surgeon would stick a tool in through the nose and get some force feedback as though they were operating. "That was the first medical thing I had ever done. Before that, I used to pass out if I had to have blood drawn. I was *so* not interested in medicine," she recalls. "But realizing how this technology could help train doctors was very motivating for me." Okamura began working with Intuitive Surgical while teaching at Johns Hopkins University, and continued after moving to Stanford.

You might wonder why anyone would put a machine between the highly skilled hands of a surgeon and their patient. The reason is actually the surgeon's hands. They're big, and that means a big hole. So does opening up the body cavity enough to let the light in. But a robotic appendage can be much smaller, and smaller holes mean less scarring and collateral damage to nearby tissue. "That's less invasive to the patient than cutting them open and having the surgeon put his big meaty hands in there," says Okamura. Plus, hands are not always the right tools for the job. Robotic tools can be designed to make motions better suited for special tasks, and also to scale their movements for extremely delicate work, so that a large maneuver on the controller becomes a much smaller one on the actuator. "Imagine you are trying to tie a doll's shoelace," Okamura says. "You could tie a person-sized shoelace on this side and a robot re-creates that on the doll."

So part of her lab's art is to re-create the tactile sensitivities and abilities of the hand without having to copy its form factor. Touch falls under the larger umbrella of somatosensation, or all of the sensations felt through the skin and muscles. This includes temperature; we have two kinds of thermoreceptors that sense when the temperature of the skin rises or falls. And as we'll see later in this chapter, we have sensors in our muscles and joints for body position. But "touch" refers specifically to the way we sense pressure. Both the skin and the dermis, the layer beneath it, are embedded with mechanoreceptors, and just as with the other senses, these are specialized to respond to different qualities of information. They can vary by receptive field—the size of the area that must be touched before they respond—as well as by the kind of touch and whether the stimulus is continuous or rapidly changing.

These receptors contain one of four types of nerve fibers, each sensitive in a different way. Merkel cell–neurite complexes, which are prominent in the fingerpads, are especially tuned for detecting fine spatial resolution, points, edges, and textures. Meissner corpuscles respond to low-frequency vibrations, like those caused by objects slipping from the hand, and help you control your grip. Pacinian corpuscles respond to higher-frequency vibrations, like those caused when a tool in your hand touches a hard surface. And finally Ruffini endings, which act in concert with each other over a large area, are sensitive to the stretching of your skin as your body moves.

CHARM Lab graduate student Sam Schorr is exploring activating these kinds of cells with something called skin stretch, or the sensation of an object moving against your skin—and this might be a useful feeling to give to a surgeon. Imagine you're holding a pen and pressing down on a desk to write, Schorr says. You immediately get an idea of how hard you are pressing, but where is that information coming from? "It seems like not very much of it is coming from your muscle activation in your arm, how hard my elbow is push-ing," he says, picking up a marker and trying it himself. "I think a lot of my perception of how hard I'm pushing is just coming from the sensation occur-ring on my fingertips, and especially your fingerpads and the skin."

To test how well skin stretch conveys force feedback information, he's running an experiment in which he's embedded a plastic coffee stirrer inside a puck of putty-colored rubber called a "tissue phantom," which mimics the feel of human organs. The setup mirrors what a surgeon might feel when palpating a hardened artery in the heart. Press on the puck with your finger, and you can just barely feel the fake artery within. Schorr takes the dish containing the fake heart tissue and places it on a rotating platform under-neath a small robotic arm. At the end of the arm is a white plastic knob about the size of a marble. This is the robotic "finger" that will probe the dish, search-ing for the hidden artery. Schorr makes a careful note of the artery's position and sets up cardboard blinders around the whole contraption. Now his test subjects can't look over to see the robot arm move.

Then he calls in his subject, another student, and sets him up at a separate workstation, where a computer monitor displays a gloomy camera feed from inside the darkened box, a close-up on the dish. This gives the subject very limited visual feedback. To his right is a robotic control arm, which termi-nates in a plastic wedge about the size and shape of a turntable stylus. The student grips the stylus between thumb and forefinger, and gently bobs it up and down. Over in the darkened box, the robot finger moves in tandem, softly pressing into the simulated flesh. Now Schorr explains his subject's task: Discover the orientation of the hidden artery.

The student will do this test dozens of times, under five conditions. In one, the stylus gives him no feedback, but a bar graph on the computer screen in-dicates how much force he's applying. In another, the stylus vibrates. In two more, he gets some degree of force feedback that lets him feel how hard he is pressing. In the skin stretch condition, a red button the size of an eraser head embedded in the stylus activates, gently pulling against his thumb as he ma-nipulates the device. (It's actually an IBM ThinkPad tactor—you might have one on your laptop.) "The idea here is that we're creating that fingerpad sen-sation," says Schorr, the push of the object against skin.

The test subject leans forward, concentrating as he bounces the robot arm up and down. Once he believes he's found the artery's angle, he draws a cor-responding line on the computer screen. In between guesses, Schorr leans in-side the darkened box and spins the dish into a new position. Once he's run this test on more subjects, he'll have an idea of how accurate skin stretch is for giving palpation feedback.

Why not just give the surgeon force feedback, creating an interface that pushes back on her hand to let her know how deeply she's depressed her instrument? The problem, says Schorr, is stability. When there are delays in force feedback, it can prompt a synching error between user and robot. If you push, and there's no reaction, you push again. Now the robot has gotten the signal to move twice, so when it does, it overshoots. Schorr demonstrates by adding a slight delay to his system. "You can see it gets these growing vibrations that go out of control," says Schorr. And sure enough, after a few seconds of wiggling the palpating finger snaps upward violently. You wouldn't want this to happen during surgery. If skin stretch turns out to be a good proxy for force feedback, it should convey sensation without risking this error loop. Okamura says her lab will try putting skin stretch buttons on the da Vinci manipulators, so that they move beneath the thumb and forefinger when the surgeon is pressing instruments against tissue. She thinks of skin stretch as a kind of sensory substitution—or really, as a "sensory subtraction"—because while it still works within the touch realm, it's not true force feedback. It's only providing part of the cue.

Another idea would render feedback more directly through an interface called haptic jamming. This project is being led by graduate student Andrew Stanley, who has two prototypes laid out on the desk in front of him—a single cell and a 12-cell version. These are soft silicone membranes filled with granular material. When you suck the air out of them, that material jams together, becoming rigid in whatever shape you want. The lab envisions them being first adopted as palpation simulators, because they can mimic the shape and hardness of different bodily objects. "For learning how to palpate soft tissues, localize hard lumps, and be able to differentiate between, say, a tumor and a fluid-filled cyst, you are really relying on your sense of touch," says Stanley. They might also be a useful way to train people to recognize the feel of skin lesions, bone structures, and the insides of wounds and how to detect foreign objects in the body, such as shrapnel.

Other labs exploring this technology have tried filling the membranes with glass beads, other small objects, and sawdust, but Stanley decided on ground coffee, because the irregularity of the bits lets them lock together well. (Uh, so, espresso grind or what? *Folgers.*) Stanley pulls over the 12-cell version, which is mounted atop a 3-D-printed plastic box, turns the vacuum on, and then individually plumps or deflates the cells, creating different shapes as some grow rigid and others relax. These are pretty big cells, about an inch square, but he'd like to shrink them down so he can create more complex shapes. "It's kind of like pixels in an image—the smaller they get, the finer the resolution, and the more pixels, the more detailed geometries you can display," Stanley says. (By early 2015, the lab was working on a 100-cell version.)

This project, part of a Department of Defense grant and a collaboration with Intelligent Automation, Inc., is designed to go inside a virtual reality display. There's a giant arcade-style black box behind Stanley, mostly empty at the moment. This, he explains, will become a virtual bedside for training medical students. The student would stand in front of the console and look down onto a screen showing the body part to be examined, while touching their hands to a platform below, which could change shape or texture using Stanley's device and those built by other collaborators. But haptic jamming might be useful for practicing surgeons, too. While a multicell display would be too complicated to put on a da Vinci manipulator, Okamura says, "you could imagine putting just a single one of those cells underneath the fingertips, and just changing softness and hardness to reflect the thing that the person is touching." Instead of substituting one tactile sensation for another, as skin stretch does, this would essentially relay it, giving the surgeon the illusion of touching flesh.

DR. WREN HAS ALMOST FINISHED WITH THE GALLBLADDER. The surgery at the Palo Alto VA hospital is in its second hour, and the mood inside the dim room is calm but upbeat. Wren calls out cheerfully to her staff; they swap tools and positions as she directs; the radio in the background plays a soft mix of opera, jazz, and Elvis Costello. Wren has finished teasing away the fat. She's applied the clips to the duct and cut through it, unmooring the gallbladder from the body. Now it's time to get rid of it.

Da Vinci users can divide up control of its many arms, a way for experienced surgeons to train others. So now that they're almost finished, Wren has her colleague Dr. Salles move onto the operating room's second console, while a third surgeon, Dr. Daphne Ly, takes over at the patient's bedside. The scrub nurse has inserted a new tool, a device that deploys a plastic baggie into the body cavity, and now the three of them are trying to wriggle the gallbladder into it. "Grab that piece of gallbladder right down there. Right here," Wren calls across the room while Salles maneuvers her tools. "Grab a nice big old bite of it." As they move the slippery organ across its equally slippery environment, Ly uses her grasper to gently shake the bag, helping them slide it in, the way you might slither wet laundry into a sack. Once it's inside, they pull the bag's drawstring, the gallbladder is ready to remove, and the robotic part of the job is over. The nursing team begins to withdraw the instruments, then pushes the entire robot against the wall, leaving the patient lying on the bed, still draped with only the port through her belly.

Even a robotic surgery requires the doctors to touch the patients directly and make judgments based on tactile feedback. At the beginning and the end—or the open and close—the surgeons work by hand. Wren had begun the operation by palpating the patient's abdomen, then plunging a finger into the incision all the way up to her third knuckle, sweeping her finger around briskly, feeling for obstructions and scar tissue. Working by hand can be incredibly physical; Wren once tore her rotator cuff during a cancer surgery. And Wren is remarkably touch oriented. Her office desk is covered with doodads that she keeps around because they feel good: a Mardi Gras necklace that she fiddles with like worry beads, a wire toy that changes into a million geometric shapes as she pushes and pulls it, a tiny hematite pig that she likes to roll between her fingers. "I'm a twitch. That's why I have all these toys," she says. "I'm very tactile." She's got a surgical instrument on her desk, too, a needle holder, and as she talks she unconsciously flips its handle around her finger like a cowgirl twirling a six-shooter, or flicks its lock mechanism open and closed. The heft of her hand tools is important to her. "This is a fine instrument," she says admiringly. "*It feels right.*"

Her relationship with the human body is remarkably physical, too, and a large part of the reason she became a surgeon. "It's like a big puzzle," she says of each patient's body. "You try to figure out what's wrong with them. And then you get to see if you're right and take care of the problem." To her, fitting that puzzle is a pleasure. "People's anatomy is beautiful on the inside, and being able to manipulate that anatomy—it's an art," she says.

So why can she function without touch when she uses a robot? Part of it is probably because she's so experienced. She's been operating since 1986, originally training to do traditional, or "open," surgeries. When laparoscopic surgeries became increasingly popular in the 1990s, she learned to do those, too. In this kind of surgery, the doctor operates through ports set into keyhole incisions, using long-stemmed instruments similar to those for robotic surgeries, but manipulating those tools by hand. Many doctors have compared laparoscopic tools to touching something by chopstick—you can feel it, but indirectly and at a distance. "There is some touch, but it's really diminished," says Wren. She stabs at her desk, imitating a long-handled instrument. You can tell if the object at the other end is hard or soft, but not *how* hard or soft, or whether it's pulsing, as an artery would. "It's a binary touch," she says. "It's not degrees of variation."

Still, her experience taught her to correlate visual cues with tactile ones. It showed her how to recognize deformation, when the tissue around a tumor moves fluidly, but the tumor itself doesn't. It taught her how to recognize where the tension is on the tissue, and the best plane on which to cut. It taught her how to know when she's pulling too hard, especially during suturing, by spotting when tissue begins to blanche. Suturing, she says, is the hardest thing for people to learn on the robot. "In the beginning, when you first start robotic suturing, people will snap the sutures because they're not feeling it," she says. "Until you learn visual cues." So touch, she thinks, might be more critical for beginners, while an experienced surgeon can mentally substitute visual cues. She's interested in the idea of restoring touch, but she wants to make sure adding the sensation back won't become a distraction to those who have learned to operate without it. "It all depends on what would it be like. What would the interface be? Would it be too much?" she asks. "Is it going to be on there all the time, or am I going to have to click it on?"

Still, she is eager to try whatever's next for robot-assisted surgery. "It's like when I was a kid and I got that first Texas Instruments calculator that cost a fortune and was this big and did four functions," she says, indicating something the size of a brick. "We're looking at first-generation tech. Where's it going to be ten years from now? It's going to be different and it's going to be better. I work pretty well with this. But I'm looking forward to what's coming out."

Once it is time to close up today's gallbladder patient, the surgeons are back to relying on their hands. They remove the port from her belly and get ready to sew. Once again Wren plunges her fingers deep inside, feeling for the fascia—the tough white connective tissue that covers muscles—to make sure they suture the right structures back together. Ly walks over to the bag containing the gallbladder and firmly palpates it with a gloved hand, counting the gallstones. Just by pressing on it and watching the flesh turn pale, she can see that there were four, the size of peas.

Wren and Salles use a tough purple suture to sew closed the fascia, then close the skin using a tan thread that will dissolve into the body, leaving barely a trace of their work. The anesthesiologist and nursing team begin waking the patient up. Wren gives her team some orders about painkillers and antibiotics, and then heads across the room to make a call. "This is Dr. Wren. How are you? I just wanted to let you know your mom's surgery went great," she says into the phone. "We're all done."

INTUITIVE SURGICAL, THE SUNNYVALE-BASED COMPANY that makes the da Vinci Surgical System, has its own mock operating rooms, extremely realistic-looking hospital bays where surgeons can practice with a robot. Simon DiMaio, a senior manager for the company's applied research division, and Anthony Jarc, a medical researcher, have set up one of these rooms for a demonstration, placing on the bed a "patient model," a torso-sized, igloo-shaped, hard plastic dome. Jarc inserts a few robotic instruments through ports built into the dome, and on the screen of the surgeon's console, we can see what's inside— rubbery pink tubing standing in for a bit of intestine.

Looking into the console's two eyepieces, I see a stereo image of the ersatz guts. The view is intensely colorful and brilliantly lit, the tubing shining as bright as Pepto-Bismol. DiMaio shows me how to grasp the manipulators with a thumb plus either the index or middle finger. They open and close like twee-zers, representing the jaws of the instrument. Little Velcro loops hold my fingers in place, so I don't have to look down at them. "Squeeze and release the grips and then the instruments will start to follow your hands," he says. On the screen, the two grippers begin to move, and I gasp because the world has magically become frictionless and gravity free. My instruments seem to have no weight at all. I move easily and without thinking about it, feeling enormously powerful, a giant poking about in a marshmallow universe. I can see that I am squeezing the pink tubing, which looks as soft as spongy honey-comb. But the only clues I have about its rigidity or how tightly I'm squeez-ing it come from watching the jaws of the grippers move together, and seeing how severely the tubing deforms. I try poking at it, and that's even more con-fusing because it hardly gives—I have no idea how forcefully I'm pressing.

But DiMaio points out that I'm not truly without touch. Because I am squeezing the grips, my fingerpads are deforming against them, providing some information about how hard I'm gripping. "But you are not feeling a force that directly maps to what you're applying with the instrument," he says. And the robot is giving me haptic feedback in other subtle ways. One aspect of somatosensation is proprioception, or the sensation of where your body parts are in space. A separate class of cells called proprioceptors within the muscles, joints, and tendons picks up information about the angles of your limbs, their movement, and changes in muscle tension. That's what lets you touch your finger to your nose with your eyes closed. Or, if you are a surgeon operating without looking at your hands, it's how you feel where they are in space. "Those hand controllers are not completely passive. They are little mini-robots," DiMaio says. "Their orientation matches the orientation of the in-strument that you see when you look inside the console." What I see matches what I feel in my hands and arms as they move, making the experience less disorienting.

DiMaio asks me to try rotating my wrist as far as it will go. Once I over-extend, the machine pushes back lightly, as the instruments reach the edge of their ranges of motion. Then DiMaio walks across the room and nudges one of the robot arms that curl over the bedside and my fake patient. As soon as he collides with it, I feel pushback from the controllers. This is a safety mechanism, warning me if I've piloted the arm into my surgical assistant, the patient, or a piece of equipment. It's important to note that these forces aren't being conveyed from the patient's body or the parts of the machine that are inside it—they are coming from the controllers themselves. But these haptic cues make the virtual world seem natural. As Okamura might say, they keep the physics real.

And realism—well, intuitiveness—was the company's mission. It was founded in 1995 as a challenger to both open and laparoscopic surgery, which had helped patients by making incisions smaller, at the price of making movement less natural for surgeons. Running all the instruments through a port causes a "fulcrum effect." To move the tool's tip right, you move your hand left. A small movement outside becomes a large one inside. "So everything is almost inverted," says Jarc. Perhaps most frustrating to surgeons, laparoscopic instrument tips are fixed rigidly to their handles, so surgeons can't bend and rotate them as they would their own wrists during an open surgery. Intuitive designed about 60 surgical instruments, most of them "wristed," or with a flexible area right before the instrument tip. And although the surgeon must still operate through a port, now that they control their instruments through a computer, the system can translate their actions in such a way that once again, right feels like right and left like left.

When the company got started, the U.S. government had already begun funding robotic surgery research, with the goal of developing mobile units for the military that could be driven into battleground areas, allowing the surgical team to stay safe by working remotely. Maneuvering a robot from a distance is called teleoperation (in this case, telesurgery), and allowing that operator to not only manipulate but also sense and communicate from a distance is called telepresence. Intuitive Surgical licensed technology from several university groups that had been involved in this early work, and redesigned a telepresence system for civilian operating rooms.

The trade-off for putting a robot in the middle of the transaction, of course, is losing the sense of touch. Allowing surgeons to feel body parts they cannot reach would require putting haptic sensors on the parts of the instruments that go inside the body. And that's a big technical challenge. They'd need to be small enough to fit on the tips of the instruments, and, thanks to sterilization requirements, either cheap enough to be disposable or tough enough to survive an autoclave.

Then, you'd need a good way to convey touch information to the surgeon. You can render gross forces by pushing back on the instrument the surgeon is holding, but there is, of course, the stability problem. And complex cues like texture and firmness require more complicated, distributed mechanisms.

"You need some kind of tactile display, not just a chunk of metal that you are gripping like a tweezer," DiMaio points out. "You need something that can stimulate your fingerpad in a way that stimulates smoothness or roughness or a lump." Haptic jamming might be a solution—like Okamura, he envisions a pad built into the manipulator. Or, he says, skin stretch might work, either as a single point, like Schorr was trying in his experiment, or as "a whole kind of forest, a little strip, little rods that stretch parts of your fingerpad."

As Wren pointed out, the trick will be ensuring that reintegrating touch information isn't distracting. It's better to omit sensory information than to risk integrating cues that don't work well, which is why Intuitive gives their users more vision than touch—current video technology is just higher fidelity than tactile technology. "If you restore a channel that is at all unpredictable or at all uncertain to the user, they'll then neglect it. It will become noise to them," says Jarc.

The da Vinci does give the surgeon a few touch-related superpowers: It can scale movement up or down, and filter out hand tremor. But for now, the superpowers on offer are mostly visual. At one point during the gallbladder surgery, Wren had asked the operating room, "You guys want to see something super cool?" and hit a button that turned everything on her screen different shades of green. She had previously injected the patient with a fluorescent agent, which made the gallbladder ducts stand out under the light of a near infrared fluorescent laser. There are two ducts that branch off from the gallbladder. By touch, they feel the same. Surgeons sometimes cut the wrong one. Seeing the junction, now clearly visible thanks to the dye, let Wren be sure she was snipping the right one. Fluorescent biomarkers can also show blood-flow, helping surgeons ensure that blood vessels or bowel sections they have reconnected are not leaking. Theoretically, they could be used to bind to cancers, showing surgeons what to cut and what to leave behind.

Okamura has ideas for haptic superpowers she would like to add. "We can put 'no-fly zones' around the very delicate organs so that the surgeon won't go in there," she says. (Another kind of surgical robot, she points out, already gives warnings to stop orthopedic surgeons from drilling too far into bone during knee surgery.) DiMaio agrees it's possible—because the robot is very accurately tracking the location of its instruments, when it gets too close to sensitive areas it could warn the surgeon with a vibration or glowing light, or simply stop moving. Perhaps, Okamura imagines, you could give surgeons "superhuman sensing capabilities" by scaling up touch feedback. Imagine doing microsurgery in the eye, where the forces are often too small to feel by hand. But what if, she says, you measured and amplified those forces, so for the surgeon, "It's like he's operating on a *giant* eye."

Or, she asks, "can the surgical robot be semi-intelligent? Can a third arm come in and help do the task that the surgeon is doing with two arms?" For example, maybe it would assist with retraction. How will the surgeon know what the extra arm is doing? "We are thinking seriously about skin stretch," says Okamura. As the robot pulls, the surgeon would feel that skin stretch, perhaps on their forearm or foot, since their fingertips are already engaged.

One of the superpowers that inspired telesurgery is, of course, the idea of operating over great distances. Theoretically, surgeons could operate on people at sea, in space, or on remote military outposts. Specialists in urban areas could attend to patients in rural areas or developing nations. Yet while that may be a very good idea, it's hard to do. DiMaio points to a blue fiber-optic cable running along the floor that carries the commands between the master console and the robot. At the moment, their systems can use a cable up to 20 meters long, which is the length they have tested and shown to be safe. But longer distances pose serious questions about latency. Remember Sam Schorr's demonstration with the palpating device: Any delay that causes the operator to move more than intended, and the system to overshoot, can start oscillations that grow out of control.

And some distances are just too long for cable. "Obviously, you are not going to have a blue fiber going from here to space. So you have to have a wireless communication," says DiMaio. "How would you make sure that that's stable and robust, and if you have a sudden sunspot that it's not going to glitch out?" Even on Earth, says Okamura, wireless isn't always reliable. "Think about how well your Internet works at home, or your cell phone," she says. "And now imagine: OK, I am going to provide surgery over this line. *Dangerous!*"

Wren, who frequently works in Africa with medical relief groups, points out that utilities glitch out all the time in low-resource areas. "If you have electricity in a steady way less than 50 percent of the time, and running water less than 50 percent of the time, there are some real issues," she says. "I can't tell you how many times I've been operating and the power goes out." And even with guaranteed steady power and telecommunications, she says, you still need actual human beings at the bedside: Someone has to insert the ports; someone has to give the anesthesia.

Still, there have been some successful long-distance surgeries. In 2001, a team of French surgeons operating in New York performed a gallbladder removal on a patient in Strasbourg, France, the first robotic surgery over transoceanic distances. They used a dedicated high-speed fiber-optic line provided by their partner, France Telecom, and the surgeons operated from inside the company's New York office, because at the time most hospitals did not have such a line. In 2003, two Canadian hospitals established the first remote telerobotics surgical service, allowing a surgeon at a teaching hospital to work with patients and nursing staff in a rural community. Both used the ZEUS Robotic Surgical System, a prior competitor to the da Vinci.

But Okamura thinks the next frontier for teleoperation is not going to be farther out, but closer in. Instead of manipulating the robot arm with your hand, it will *be* your hand. "A prosthetic arm is actually a teleoperated robot," she says. "A human brain would control its movements. You would sense feedback and you would give it back. So it would ideally be a teleoperated robot, but brain controlled, rather than through a master manipulator." To see where this idea is headed, we have to leave behind the da Vinci, which is already a part of routine practice at hundreds of hospitals, and move into a much newer area of basic research. And that means going back to Stanford.

SERGEY STAVISKY AND JONATHAN KAO ARE SITTING at their rig in the basement of a Stanford life sciences building, watching Monkey R through an infrared video camera. Monkey R is a rhesus macaque, and he is sitting in a special chair that allows him to move one hand while the other extends through a little tube. Monkey R is facing a virtual reality screen, which shows him a blue dot floating in 3-D space before him. That's his target.

Monkey R's job is to move a smaller gray dot, the cursor, onto the target and hold it there. If he does, he gets a drop of juice. As Monkey R reaches toward the screen, his left hand, going through the tube, rests, while the right one moves freely as he reaches through space. But the monkey's not controlling the cursor with his hand. He is controlling it with his brain.

This monkey has two 100-electrode arrays implanted in his brain; you can just see the cables that connect to a port on his head during experiments. These arrays are tiny, just four millimeters square. They look like little hairbrushes, each bristle only a millimeter long. The arrays read out signals from Monkey R's primary and premotor cortex as he moves his arm. A computer display shows the experimenters a grid with 100 boxes. Inside each, a white squiggle indicates the activity on one electrode. Over the lab's speakers, we can also hear it, a whispery static. "Every crackle is that neuron firing off an action potential," says Stavisky. "It's traveling down the axon down to the next neuron or down towards the muscles."

Stavisky is a graduate student in neuroscience and Kao in electrical engineering; they both work in the lab of Dr. Krishna Shenoy, one of the nation's leaders in neuroprosthetics research. Theirs is primarily a motor prosthetics lab, more focused on learning how to read out movement signals than to write in touch cues. But Shenoy and Okamura have just begun to collaborate, given their common interest in developing better prosthetic limbs, which would allow people to not only move, but feel. "We absolutely need to—in preclinical trials and clinical trials—move on to bestowing on our subjects the sense of touch. We *have to*," says Shenoy. Touch has practical implications for anyone actually trying to use a prosthetic limb. For example, he says, a user trying to pick up a coffee cup needs tactile feedback—not just visual. "If we reach out and try to pick up a cup of coffee and we squeeze too tightly and crush the cup, or not tightly enough and let it slip, we know that that's not very useful," Shenoy says. "When you are wrapping your fingers around the coffee cup, you don't have a great view of that. Your vision is very limited in that particular circumstance. You and I both rely very heavily on the pressure we feel on our fingers, that sense of touch."

In addition, Okamura points out, touch would enhance the user's experience, because it's an important way to feel connected to other people. "Ever since I have had kids, I have an even better appreciation for the sense of touch," Okamura says. "Loving physical contact between people is so crucial to us being human and developing correctly." When she began looking into neuroprosthetics, she was particularly struck by studies finding that users desired to sense temperature, "so they can feel the warmth of their loved one's hand."

The prosthetic limb is an incredibly practical use for a brain-machine interface, and a research field that is more well established than the inner voice translator imagined by the Gallant Lab. The Shenoy Lab is one of many in the United States doing cutting-edge research. Other notable labs are at UC Berkeley, Duke University, and Brown University, which is working in conjunction with Massachusetts General Hospital on the BrainGate FDA clinical trial to develop implantable interfaces, much like the ones Monkey R is wearing, for people with paralysis. So far, most of the field has focused on reading out, finding a way to translate motor cortex activity into signals that can drive the limb. The next challenge is writing in, and marrying motor read-out and sensory write-in functions together in real time.

It's not nearly as hard to put sensors on a robotic limb as it is to put them inside a surgical robot. There aren't the same space and sterility constraints. "The tricky part is, well, how do you get that information back into the brain?" asks Shenoy. Instead of conveying sensory information to, say, a joystick and then on to the hand, which already has its own mechanoreceptors and an information pathway to the brain, now you have to figure out the touch system's electrical language, just as Second Sight did with vision for the retinal implant. As with vision, it's one thing to generate a binary sense of touch or no touch, light or no light. But it's harder to write in shades of information, like a soft squeeze versus a hard one. In early experiments with stimulating the somatosensory areas of animals' brains, people trained the animals to respond when they felt a stimulus. "What becomes a little trickier is knowing the quality of that percept: What are they really sensing?" asks Shenoy. "And you can't really ask a monkey that question."

But you can listen in on a monkey's brain activity to see what it's doing when the monkey is moving naturally and getting touch feedback. Stavisky and Kao's work focuses on how Monkey R learns to adapt his control of the prosthetic as they give the animal feedback simulating resistance or weight. Stavisky can change the speed of the monkey's cursor so it appears to have a different "mass." When it's purple, it's slow or "heavy." The blue one is medium. When it's orange, it's fast or "light." Stavisky is randomly changing the mass on each trial, so each time the monkey reaches out, the cursor "feels" different to him, and he has to adjust his movement appropriately.

In this experiment, they're studying internal models, or how your brain learns what to expect when you try to move something with your arm. This will be important to prosthetic wearers trying to interact with the physical world. Imagine that coffee cup again. "If I know it's heavy, I'm going to apply a bunch of force," says Stavisky. "If I know it's going to be light, I'll have a more delicate touch." Now imagine that the cup you expect to be heavy is actually empty. You pick it up too hard. Maybe it flies away and you spill dregs everywhere. The same thing essentially happens to Monkey R when Stavisky drops the mass on the cursor. Stavisky points to the spheres moving on the screen: "There it went from a medium to a light one, and you see he kind of shot too far."

But Stavisky wants to see if Monkey R can adapt his internal model, to learn to control the brain-machine interface as he would his own arm. So now Stavisky taps some keys, and instead of the cursor switching randomly, it holds steady at purple, or "heavy." Can the monkey adjust, and will he perform better if he does? In the data they've collected so far, it looks like the monkey can: The monkey overshoots by about twice as much when the cursor changes randomly than he does when the cursor stays steady. "So this makes us more optimistic that we'll be able to control robotic arms in the real world," Stavisky says, where people need to be able to adapt to the feedback they get from objects.

Across the hall, neuroscience graduate student Dan O'Shea is doing an experiment that pushes into the touch world even further. His setup is similar. A different macaque, Monkey P, is seated in a chair, facing a screen and a straw of juice for rewards. At this point, Monkey P is in training and has no electrodes implanted in his brain. He is controlling the cursor on the screen by using his hand to move a very fancy joystick. The monkey's job is to pilot the cursor up a simple pathway flanked by red walls he must avoid. Sometimes the channel is straight. Sometimes it zigzags left, sometimes right. The monkey's cursor starts at the bottom of the channel, and he must steer it to a target at the top. But sometimes, the joystick will unexpectedly jog his hand right or left. If the sudden movement makes Monkey P run into a wall, the motors in the joystick resist, feeling to him "like he's collided with a very soft springy block," says O'Shea. If the monkey hits a wall, no reward.

O'Shea is studying how proprioception guides motor control, specifically how that feedback influences the way the monkey adjusts his reach after he gets the signal he has veered off course. Now, Monkey P already has visual feedback—he can see when he hits a wall. But for a prosthetics user who wants to do complex or precise tasks, vision is just too slow, O'Shea says. If they have to watch their limb to know what it's doing, there will be a lag before they can react. Proprioceptive feedback is faster. Once Monkey P has been trained, he'll be implanted with an electrode array, so O'Shea can study his brain activity when his hand is "perturbed" by the feedback from the joystick. Being able to write in similar collision information to a prosthetic user would help them control their limb, O'Shea says, because they'd get "this nice tight feedback loop." They act, he says, and "the consequences in the real world get reflected back into the brain very quickly and transparently."

A tight feedback loop is the ideal. As with robotics, brain-machine interfaces also face a latency challenge if user and machine get knocked out of synch. Kao calls up a video from another experiment, in which a monkey is using a brain-controlled cursor to type. The monkey sees an on-screen "keyboard" made of yellow dots, each containing a different letter. A target indicates where to move the cursor, and as the monkey lands it on each dot, letters pop up on the screen. "Hello world," they begin to spell out, as the monkey follows the cues.

This is exactly the kind of program that would be useful to human brain–machine interface users, people who can't speak or move but could thought-control a machine. But there are still many ways to make syncing errors. "What happens if the cursor control is so bad that while you are moving towards a new key you accidentally dwell on another key and you communicate the wrong letter?" asks Kao. "Then you'll have to hit backspace and all of a sudden it's much more complicated." If your accuracy was 50 percent, he points out, you'd spend so much time backspacing "you wouldn't be able to write anything."

But so far, this monkey isn't doing half bad, managing about ten words per minute. "I am controlling this cursor with my mind," read the words appearing on the screen. "I have been using this model of decoder every day for the past month. I wonder if you had a million Shakespeares, could they write like a monkey?"

Shenoy's lab is participating in the multisite FDA clinical trial of the Brain-Gate2 Neural Interface System, a first-generation implant for people, which is similar to the ones used in the monkeys. The Stanford team works with a woman paralyzed by Lou Gehrig's disease who has been implanted with the device, and they ask her to perform very similar tasks. "Just thinking about moving her arm or her fingers," says Shenoy, "she is able to move a computer cursor on a screen and type out messages to her loved ones, or to us, or to her physicians." While he can't give too many details—the trial is ongoing—he's pleased with the team's progress. "It's really exciting times," he says enthusiastically. "It's showing that what we're learning in the lab does translate."

But reading out an intention to move is easier than writing touch information into the brain. Motor activity involves millions of neurons, but reading out information from a just few hundred cells is enough to roughly correlate brain activity with intended movement. "Whereas to write stuff in, you probably have to actually write it in *correctly*," says Stavisky. "Correlation may not be good enough." You might need hundreds of thousands of neurons to receive the specific input pattern they would have gotten if the body was truly touching something.

And as Shenoy points out, writing in degrees of sensation, not just binary touch, means arrays with more electrodes, and more electrodes require more current in the brain. That's a problem because current can move uncontrollably. It doesn't always stimulate the cells—or the number of cells—you want it to. "If you imagine a piano keyboard, where each key on the keyboard is a different neuron, electrical current is sort of like taking a couple of foot-long two-by-fours and clunking it down on the keyboard," says Shenoy. "If I want to stimulate neurons harder, I can hit the keyboard harder with that two-by-four. Or wider, I can maybe move it around while I'm hitting it. But you can't really play one single note, right? And you certainly couldn't play in a pattern of notes." (If you'll recall, Second Sight anticipates a similarly diffuse "cloud effect" from putting too many electrodes in a retinal implant.)

But there is, Shenoy thinks, a solution, one that will allow finer degrees of sensory input, and that is to replace the electrode array with optogenetics. Unlike current, optogenetics would allow you to hit only the keys you want. This is much like the method Sheila Nirenberg has proposed as an alternative to the retinal implant, stimulating gene-altered cells by shining a specific light at them. Her model proposes altering genes in the eye and beaming light at them through a miniprojector mounted in glasses. In Shenoy's concept, the fiber optics would be implanted in the brain and directly stimulate the desired neurons. Investigating that, he believes, will be the field's next chapter.

And if you want to flip several chapters ahead, it's likely that the story of writing in and reading out, which began in the periphery, with the sensory organs and limbs, ends with the brain. Shenoy points out that in the United States there are some 6 million people with paralysis, the primary user base for brain-machine interfaces to drive prosthetic limbs. But many more people will soon be facing the diseases of aging, maladies of the brain itself, like Alzheimer's and other dementias. "The coming storm of neurodegenerative, neurodebilitative diseases that often go along with aging in America will really require a deeper understanding of what different brain areas do, and then some way to treat it," Shenoy says. "I mean, nothing is more depressing than going to see your neurologist, because they will tell you exactly what's wrong with you and then can't do much to help you."

But Shenoy thinks there will be help soon, and it will be because of this vast, multipurpose effort to parse the brain's language. "Once you have the capability of reading information out of the brain and writing it in to the brain, then you can think of all sorts of things," he says. For instance, imagine that someone has had a stroke, and cells in one part of their brain died. If you know enough about what that damaged area normally does, you could mimic its function with an implant. And if you know how that area fits into the brain's linguistic relay, you could read out information from just before the stroke area, run it into the implant, and then write the transformed information back in just beyond the damaged spot. "You patch around it," Shenoy says. "So it's a brain prosthesis," just like a vision, hearing, or motor prosthesis.

The retinal implant, stimulus reconstruction, the robotic limb—these are early attempts at translating the language of the black box. But they are also a crack at daylighting the circuitry inside. Scientists have a powerful motivation now: to remedy not only blindness, strokes, and paralysis, but an entirely new category of age-related cognitive disorders that very few people in previous generations lived long enough to experience, and that we have few tools to treat.

"On the near-term horizon, we are reading out information and we are writing in information. But where is it all headed?" Shenoy asks thoughtfully. "It's headed to a standardized way in which to read and write from the brain such that we can address a myriad of diseases and injuries. And that really is exciting. It is something that pharmacologicals cannot do, and surgery cannot do. It is a new class of medicine. And that class of medicine is the brain prosthesis."

PART TWO

....

Metasensory Perception

SIX

....

Time

ALEXANDER ROSE LEANS BACK in the café chair, and in a soft, steady voice, begins to tell the story of how you might visit the clock:

You and your fellow adventurers would drive into the desert, somewhere past the town of Van Horn, Texas. Around sunrise, you would begin your hike across the valley floor, aiming for the rock cliffs, walking a canyon path that gets narrower and narrower, until you arrive at the cliffs' base. You'd find a curious opening that looks like it could be either geological or man-made, and if you ventured inside, after a few hundred feet of dim cave pathway, you'd come upon a door. If you were brave enough to pass through, you'd see that the cavern walls had changed from rough limestone to hewn rock, that you'd moved from the natural world into a mysterious built one.

"If you looked up," Rose says, "you'd see light filtering through from the surface, even though you are 500 feet underground. You'd be seeing it filtering through what looks like machinery over your head. And there's a spiral staircase and a shaft. So you'd start walking up the spiral staircase." This machinery, an enormous, carefully wrought column of stainless steel instrumentation inserted into the mountain bore, will be the clock, meant to last for 10,000 years, a monument—and a challenge to—the human perception of time.

Rose is the executive director of The Long Now Foundation, the group that is building the clock, under construction now with a launch date in the undetermined future. But the foundation members have already richly envisioned that otherworldly ascent up a twisted staircase. As he sips coffee in a sort of oddly timeless café—1940s pop standards wafting through the background clatter, classic neon on the walls—near the foundation's San Francisco office, somewhere out in Texas, a giant diamond-edged saw is carving limestone slices from the bore, creating the central stairway that will wrap around the clockwork. In Seattle and in California's North Bay, engineers are machining the clock's internal mechanisms, trying to fabricate parts sturdy enough to stand the test of time. A heck of a lot of time.

The clock is the brainchild of star programmer Danny Hillis, who conceived it, Rose says, "as a kind of antidote to what he had spent most of his life working on, which was building the very fastest supercomputers in the world." The problem, Hillis felt, was that the cultural emphasis on speed was prompting people to sacrifice the good of the future for the needs of the present. As a species, we had become too focused on short-term time spans like electoral cycles or fashion seasons. "His fear was that there were certain problems that could only be solved on a much longer timescale," says Rose, "things like climate change or hunger or education. These are not things you solve in a four-year election horizon." The clock would be a counterexample, Rose says, essentially "the slowest computer in the world," a behemoth that would force you to see yourself not at the forward edge of time's progress, but at just one point along its way. "The original kind of poetic version of it was a clock that ticked once a year and bonged once a century and the cuckoo would come out once a millennium," says Rose.

Rose, a lanky industrial designer with a thoughtful, deliberate manner, was The Long Now's first hire. The nonprofit was established in 1996—or rather, 01996. (They've already converted to a five-digit year.) The phrase "long now" was borrowed from British musician—and Long Now board member—Brian Eno, who realized upon moving to the United States in the 1970s that "now" in New York City meant something a good deal faster than it had in England. "He called having a larger time sense the 'long now,'" says Rose. "And so we took that and stretched it out even further to mean really the last 10,000 years and the next 10,000." They chose 10,000 as a civilization-scale time marker. You can count it backward to the beginning of the anthropocene, or the rise of agriculture around 8000 BC, when people began to have a discernible effect on their environment. And now the clock will count it forward.

Rose continues describing what you'll see if you climb the clock's stairs: "You would pass lots of machinery, all of it still. Some places you'd get a chance to actually wind it up." The staircase will lead to a main room with a display of dials, which will show the date when people last visited—maybe yesterday, maybe a thousand years ago. In case people make this visit so far in the deep future that they no longer understand the Gregorian calendar, this display will also read time out in the position of the sun, moon, and stars.

Visitors looking at the dials, says Rose, might notice another curious thing: "There is a pendulum still ticking away, even though the dials haven't moved in a thousand years." There would be an obvious-looking hand winder, and if you chose to crank it, he says, "you'd see all the astronomic dials move and all the dates move and update. And they would stop when they hit now."

The clock will have two power sources: hand-winding and a thermal mechanism that harnesses the temperature differential between day and night. Extra energy will be stored in a weight system, enough power to keep time for 100 years without sun or people. When the clock is fully wound, it will chime at solar noon, a unique peal of ten bells each day for 10,000 years. (Or at least very nearly. Eno designed the sequence, which will produce almost enough combinations for the 3.65 million days of the clock's life, assuming it won't ring on the days when it has no visitors.)

As the chamber spirals around the clockworks, Rose says, it will get tighter and tighter, until you finally emerge in a cathartic breakthrough at the top of a cliff. "Then you have this big 270-degree view of the high West Texas and New Mexican desert," he says. But more importantly, after your close experience with this massive time sentinel, you and your friends are "hopefully somewhat changed by the experience, and travel home together, talking about it and thinking about it."

And thinking about time is complex. The first five chapters of this book were about the senses, each a distinct modality that channels information to the brain through a specific nervous system pathway. But perception also includes metasensory, or polysensory, experiences, which draw information from multiple sense organs. There is no one time-telling organ, no time cortex or lobe; indeed, the neural function of timekeeping is likely distributed throughout the brain. And we use many of our senses together to gauge time. Imagine a herd of wild horses running across that Texas flat. You can see time in the arc of their approach, in how they seem to grow bigger as they get closer. You can feel time, if you lay your hand on the trembling ground. Thanks to the Doppler effect, you can hear time as they thunder past. And your brain, ultimately, edits time—one of its most fascinating perceptual mechanisms. It must, because time comes to us through these organs at varying speeds. It syncs these inputs to produce a coherent experience for you, so that the horses and their noise and the vibration from their hooves all seem to arrive at the same moment.

Time is also a cultural phenomenon, measured through man-made devices and used to modulate our behavior, making it perhaps the ultimate commingling of hard and soft biohacking forces. We've taken time cues from natural cycles both inside and outside of our bodies: our sleep/wake cycle, the movement of planets and stars, the change of tides and seasons. We've translated these into counting mechanisms, from sundials to calendars to clocks, of which The Long Now project is merely an enormous example. But as we'll see, it's not the first major attempt to calibrate our time perception as a species, to coordinate our lives as a human society. So if you want to understand time, you have to examine it on more than one level. What kind of time you perceive really depends on what kind of clock you are reading.

WHILE THE 10,000 YEAR CLOCK WILL COUNT THE LARGE and slow, Dr. Dean Buonomano is looking for time in the small and fast. Buonomano is a neurobiologist at UC Los Angeles who specializes in time on the millisecond scale, and he's trying to figure out how your brain computes time, despite its apparent lack of a central counter.

We don't have a specific time-sensing organ, Buonomano says, because there isn't a physical property of time to measure, the way the eye reacts to photons, the tongue and nose to chemicals, the ear to vibrations or the skin to pressure. "I don't think it's that surprising that we don't have a specific sense of time any more than we have a specific sensory device to measure space," he says, "because space is universal and time is universal, meaning they are fundamental dimensions." Just as you can perceive time through many senses, he points out, you can do the same with space: You can hear it in an echo, see it thanks to the depth perception provided by binocular vision, and tell the size and distance of objects by their touch on your skin.

"So it gets to this very fascinating question: What is time to begin with?" asks Buonomano. And the answer, he thinks, is that time is a measure of how much the world around us changes. It's a mechanism that simple living organisms evolved to tell where they were in the day's dark/light cycle, and that more complex animals need so they can anticipate what others will do, and sync their activities with events in a constantly changing environment. Bees need to know when to return to a flower for a new profusion of nectar. Birds need to know when it's the season to migrate. Gazelles need to know how fast the cheetah is coming at them.

Different species require different levels of time-telling sophistication, Buonomano says, so evolution came up with separate solutions "on an as-needed basis." Plants and single-celled organisms just need to discern the alternation between dark and light; you don't even need a brain to do that. But animals (like humans) that vocalize need a highly tuned sense of time, because speech is temporal. So is fine motor control.

To be clear: No one is sure yet how our brains tell time. In fact, there might be more than one way, each perhaps telling time at different scales or for different functions. An older theory, dubbed the "pacemaker-accumulator model," supposed that the brain essentially had an internal clock that counted pulses. But the idea came under fire during the last decade, as scientists argued that a timing function was likely more distributed. Psychologist Warren Meck at Duke University, who once supported this model, is now advancing a new proposal with his colleagues known as the "striatal-beat frequency model." They argue that for time in the seconds-to-minutes range, a part of the brain called the striatum serves as its "core timer," where specialized neurons—each their own smaller timers—synchronize the oscillatory activity of cells throughout the brain. In another model, proposed by psychologist Richard Ivry at UC Berkeley and his colleagues, the cerebellum regulates timing for skilled motor function and sensory tasks that require fine timescales, using learning mechanisms to optimize the timing between inputs and outputs, without an explicit timer such as a pacemaker.

Buonomano's lab argues that there is no counter or master timing area, but rather a dynamic process that's distributed throughout many parts of the brain. "Telling time is so important for human behavior, for understanding the world, for anticipating events in the world, for motor coordination and for making sense of sensory stimuli that it doesn't make sense to have one clock within our brain," he says. In fact, he continues, "one of the emerging notions is that the brain tells time in a way that's very different from what we conventionally think of as a clock." Most clocks count the ticks of an oscillator. For a grandfather clock, that's the pendulum. For an atomic clock, it's the cycling of cesium atoms. But the brain, at least for the short timescales between a few seconds or milliseconds, seems not to—this is the timescale you're operating on when you decipher speech or react to a red light.

Instead, Buonomano thinks, the brain tells time through neural dynamics, or the activity patterns of interconnected neurons. If you give them an input, they respond, influencing one another and setting off a chain reaction until the process runs out and the system resets. It's sort of like throwing a pebble into a pond, he says, and using the dynamics of that fluid to tell time. The stone's impact creates a series of ripples, and by watching how far the ripples have traveled and how long it takes for them to fade, you can tell how long it's been since you threw the pebble in. So the brain might read time from the activity pattern produced by a network of neurons (the pond) after receiving sensory input (the pebble) and then resetting to their normal state (the ripple fading).

"Then the question is, well, where are those circuits?" Buonomano asks. There's no one answer yet: He agrees that the cerebellum, basal ganglia, and several sensory cortices, including the visual and auditory ones, may all be involved in timing, albeit on different scales. And he's careful to say that while he thinks his lab's "state-dependent network model" is likely the primary mechanism for telling time, it's probably not the only one. For example, for mammals, the suprachiasmatic nucleus in the hypothalamus controls circadian rhythms, including when we wake and sleep, using light information from specialized photoreceptors in the eyes. For time-telling on the scale of minutes or more, the function becomes intertwined with memory and likely involves the prefrontal cortex. There might be several mechanisms, which evolved at separate periods in the brain's development to address different needs. Meck adds that these models "aren't mutually exclusive and in fact can readily be viewed as complementing each other," with Buonomano's network model accounting for sensory-specific timing in the millisecond range and the striatal beat-frequency model working in the seconds-to-minutes range, and other circuits perhaps coordinating between them.

Buonomano's lab focuses on time-telling in the brain's sensory areas, particularly hearing, so he runs me through an audio experiment illustrating his argument's central theme. He puts a small laptop in front of me, and explains that it's going to emit two pairs of tones, a total of four. My job is to figure out which pair has the longest delay between them. The computer beeps twice. Then it beeps twice again in a way that seems to take slightly longer. I hit a button indicating the second pair, and the word "WRONG" pops up in red on the screen.

I end up seeing "WRONG" a lot as I play through 60 sets. The test gets harder when I get the answers right and easier when I flub them. And by "harder," I mean that the pairs of beeps seem more and more alike, making it tougher to judge which had the longer lag between them. Then we do the whole test again, with me judging between 60 more sets of paired beeps. To me, both rounds seem equally difficult; my accuracy hovers around 76 percent.

But there was a trick, Buonomano reveals. Each time, one pair of beeps was 100 milliseconds apart; the other was slightly more or less. To my neurons, whenever I heard that first pair of beeps, it was just like throwing a stone into a pond—the system started to process a reaction. When the second pair came along, like throwing another pebble into already-moving water, they had to start reacting *again*.

The twist was that the lab was varying not just the time between the beeps in each pair, but the interval *between* the pairs. In the first phase, the interval was 250 milliseconds; in the second it was 750. More time between the inputs should have made me better at my task, says Buonomano, because "the network had more time to reset, to settle back down." But not giving the system enough time to reset should have done the opposite, because the patterns created by each set of beeps would have interfered with each other.

"If you throw in two pebbles, and then you wait for the pond to settle down and throw in two more pebbles, that's much easier than if you throw in two pebbles quickly," he says.

And the difference shows in my performance. Buonomano calls up my results: In the second round, with the longer interval, my discrimination threshold was 15 milliseconds—on average, I could tell the difference between a 100- and a 115-millisecond lag between paired beeps. That's pretty normal for a first-timer, he says. But in the first round, with the shorter interval, my threshold was just over 29 milliseconds, he tells me, "so it was almost twice as hard for you." And there's the clue that time-telling relies on a dynamic system: When you interrupt it, it suffers.

But while there was a neural difference, the timing *felt* the same to me. Telling time on a neural level is not the same as our subjective sense of time, or what Buonomano calls "the construct that our brain is making and contributing to a nice narrative of the world around us." Here, your brain is doing some serious fudging. The brain constantly does what some call "backward editing," or syncing multisensory inputs to create the illusion of a continuous, linear experience.

This is not unique to time; cognitive scientist Daniel Dennett famously advanced the concept of continuous editing and "multiple drafts" for all perception. But time lets you play with this idea in some amusing ways, as Baylor College of Medicine neuroscientist David Eagleman, whose lab studies time and other perceptual functions, has amply illustrated. His lab has done experiments to see whether manipulating the delay between two events— pressing a button and a light flashing—will make them appear to switch order. (Answer: Yes.) They studied whether fear actually "slows" time by testing whether people who were falling off a tower were better at reading numbers rapidly flashing at them than people who were standing safely on the ground. (Answer: No.)

In his famous 2009 essay, "Brain Time," Eagleman asked why the brain would risk a potentially life-threatening perceptual delay in order to accommodate its slowest-arriving sensory feed. After all, he writes, "an animal does not want to live too far in the past." Ultimately, he argued, at least for conscious processes and information needed for cognitive functions, the brain prioritizes quality over speed. The brain's "carefully refined picture of what just happened is all it will have to work with later," he wrote, "so it had better invest the time."

Buonomano himself has written extensively on this topic, including a 2011 book called *Brain Bugs* about its editorial quirks, and one of his favorite time-related bugs has to do with saccade movements, which as you'll recall are tiny jerks of the eye (or of Dean Lloyd's entire head). If you look closely at someone else's eyes, you can see them doing this, Buonomano says, but if you watch yourself in a mirror, you won't, "because your brain is editing out the time that you are moving your own eyes, so you don't see a big blur rushing before you." You can also catch these edits with speech. Imagine hearing the sentences "The mouse I found was broken" or "The mouse I found was dead," he suggests. The last word changes the meaning of "mouse," so your brain delays interpretation until after you have heard it. That delay is worthwhile because what your brain cares about in this instance is not the raw information—the air vibrations that are creating the sounds—but its meaning.

So just as with the five senses, when telling time, the brain takes in and filters much more information than it passes on to consciousness. Your brain can *use* time without you *feeling* it. The editor shapes the raw work; you only perceive it when there's a draft worth reading.

THE LONG NOW'S CLOCK HAS A FIRST DRAFT, TOO. It's on display at the Science Museum in London, where it caps a collection called "Making the Modern World," 250 years' worth of firsts: the oldest steam locomotive, the first sewing machine, the Apple I computer in its wooden case. At the entryway is a pendulum clock from 1780 built for a royal observatory, a symbol of the eighteenth-century move to link the study of space with the measurement of time. And at the other end, where the present bleeds off into the future, is the 10,000 Year Clock, or at least a much smaller prototype.

"We're a museum, so we're in the time business," says David Rooney as he stands before its case. "We wanted to start and finish the gallery with a clock." Rooney's job title is "Curator of Time," which sounds like it might belong to some kind of excessively beardy wizard. In fact, Rooney is a charming, smartly dressed young historian sporting, perhaps not surprisingly, a very stylish watch. He was the clock's first guardian, winding it diligently for the first two years after it arrived at the museum in 2000. (They no longer keep it running daily, although it did help The Long Now address technical issues for their future designs.)

As the museum was planning the exhibit hall, Rooney fell in love with the 10,000 Year Clock and the idea of reframing time perception. He cites the work of writer Stewart Brand—an influential Long Now board member—in the 1970s to get NASA to publicly release its photographs of Earth as seen from space. Those images "became symbols of a nascent environmental movement," says Rooney, "pushing back our spatial horizon, seeing the world as one, seeing it borderless." So today, Rooney asks, "How do we push back *tem-*

poral horizons and think further into the future, so that long-term thinking becomes commonplace and ordinary, rather than radical?"

Actually seeing a clock might help people start talking about time, The Long Now founders felt, so not long after officially setting the prototype in motion at midnight on New Year's Eve, 1999—the clock bonged twice to mark the tick-over to 2000—they shipped it across the Atlantic. The design process for the final clock is still ongoing, so the pieces being fabricated in Seattle and Sausalito aren't *exactly* like the ones in the case, but the prototype is a gauge of where they're headed.

The central figure of the elegant, somewhat Victorian, prototype is a binary mechanical adder, a series of wheels or "bit adder" gears programmed using bit pin levers. Imagine a stack of spiky metal pancakes that slowly rotate as a column, and you've got the idea. Underneath is a torsional pendulum with a sphere on each of its three protruding arms, which twists back and forth like the pendulum in an old-fashioned anniversary clock. In the prototype, the clock is powered by weights that fall through tall columns on either side; these are what Rooney had to wind every day. (In the final version, the parts will be linear to fit within the mountain shaft, so the weight system will be underneath the rest of the clock. It will also have parts not included in the prototype, like a solar synchronizer that registers each noon, and the thermal power apparatus.) At the prototype's top is a black disk showing the star field visible in the sky; it's surrounded by a series of intricately wrought gold and silver dials that mark the position of the sun and moon, plus the year and century in the Gregorian calendar. "That will take us all the way up to 11999," says Rooney.

The London version doesn't have chimes, but these will function as their own subtle clock-within-a-clock. "It's built on bells, which is about as ancient in terms of clocks as you can get," says Rooney, noting that medieval clocks were essentially mechanical bell-ringers that called people to prayer. (In fact, he points out, the word "clock" derives from words for "bell.") Because each day's peal is unique, if you know the algorithm that generates the chime sequence, he says, "you can work backwards and find out when the clock started. Or you can start now and you can work out what the tune will be on any day in the future. So it acts like a calendar—it helps you track those 10,000 years, but in a way that's just exquisite."

Rooney's focus is on how we perceive time culturally, not neurologically—how we've measured, standardized, and distributed time to organize civil society. But, like Buonomano, he points out that time and space are deeply intertwined. People have long used space to locate themselves in time, and vice versa. We often measure distance by the time it takes to make a journey, and use words like "ahead" or "behind" to describe the order of events.

Indeed, one of the driving forces behind the development of exact time measurement was the need to measure space. Maritime navigation required sailors to precisely pinpoint their position on the Earth's surface. "Are you going to crash into rocks, or are you going to starve your crew to death because you thought you were closer to land than you were?" Rooney asks. "And the solution to that spatial problem was temporal—it was the development of accurate clocks." Today, he notes, we still do it. When we use GPS, we track space with time: "It's tiny differences in time from the signals you get from the different satellites, which are all at slightly different locations, which enable you to construct your position. So it's overland navigation on the surface of the Earth using clocks."

Another driver was the need to synchronize group activity. "We've lived industrial lives in one way or another for thousands of years," Rooney says, "and if you are going to carry out labor, if you are going to be a farmer, if you're going to run a town, then you need to divide the day up." Sunrise, noon, and sunset aren't enough—you need a finer grain to coordinate workers and pay people according to time. "And that's not a modern phenomenon," he adds, pointing out that you can find Roman writers well before the Industrial Revolution grousing about the sundial. "We've always been time-bound," Rooney says. "We've always been, to an extent, ruled by timekeeping technologies that divide the day." In the museum gallery upstairs, there are timekeeping mechanisms going back to AD 550, from pocket-sized sundials to devices that used falling sand, water, or lamp oil to measure the hours.

But the Industrial Revolution definitely tightened time up. In the nineteenth century, international trade spread by sea, and industrial society brought with it railways and factories, all driven by schedules. Time, until then, had been fairly local. Cities maintained public clocks, but these were not synchronized. "High noon" was whenever the sun was directly overhead, but noon in one city could be different than in another a few hundred miles away. This was bad news for anyone trying to run a train system. To coordinate travel between distant cities, says Rooney, "then you need to know *whose* time you're talking about." So in 1840, England's Great Western Railway established "railway time," the first effort to sync local times within agreed-upon time zones. They chose London time as their "local," which is based on Greenwich Mean Time. (Greenwich Mean Time is itself based on when the sun is over the Greenwich meridian, one of the longitudinal lines that cut from the North to the South Pole.)

Just as cities had local times, countries used their own meridians. "It's fine for every country to have their own, and that's how life was for hundreds of years," says Rooney, but "as the world became global, it became prudent to choose one, which would be the prime meridian. And a conference in Washington, D.C., was held over a couple of weeks in 1884 to try and choose whose it's going to be. Washington? Paris? Berlin? Greenwich?"

Greenwich won, thanks to Britain's history as a maritime power. London's chart makers, like the railway, had used their local meridian. And because they were incredibly prolific, most of the world's ships were using it, too. While adopting the prime meridian isn't mandatory, Rooney says, over the next several decades, "most countries fell into line with a time based on an integer number of hours from Greenwich." There are still oddities, he adds: India is five and a half hours off of Greenwich Mean Time. Nepal is five and three quarters. Australia has a half-hour zone.

Once you have a standard time, you have to let people know what it is. City clocks had done that job locally, and telegraphs and telephones allowed time signaling over international distances. But broadcasting made it truly universal. Rooney walks over to an enormous blue cabinet flecked with dials, which turns out to be one of the world's most famous timing devices. This machine produced the BBC's "Greenwich pips"—five short tones, one long— that marked the top of the hour for radio audiences, disseminating Greenwich Mean Time to the world. The BBC first sounded the pips in 1924, and continues them today, allowing everyone on the planet to synchronize their watches.

But that only takes care of the watch's hour and minute hands. There's still the question of how the world learned to agree on the second. So for a moment let's leave London's Science Museum, and stop in on perhaps the most massive experiment ever in public timekeeping: the atomic clock.

THE NATIONAL INSTITUTE OF STANDARDS AND TECHNOLOGY, or NIST, lab in Boulder, Colorado, is a series of shiny buildings slung beneath a particularly breathtaking snowy alpine mountaintop. This federal agency has a hand in the way people in the United States measure *everything*. Need to make sure a gas station is pumping out exactly a gallon for each one charged? State inspectors use a special NIST container. Need to know how much protein is in your peanut butter so you can put the right information on its label? NIST makes "standard reference material peanut butter" that you can use to check your mass spectrometer readings. Need to calibrate the power of your laser, or your grocery store scale? They've got you covered. And one of the things

NIST measures is time, hyperaccurately, down to the level of parts in ten to the 16th power (officially: 3×10^{-16}).

NIST public outreach coordinator Jim Burrus and engineer John Lowe are standing in a long hallway, basically surrounded by atomic clocks. To their right is the room with the primary standard clock, the one that sets the official time signal for the nation, and whose readout is pooled with atomic clocks around the world to create Coordinated Universal Time based on "atomic time." (This is the standard used by the scientific and telecommunications communities, and it's essentially the same as Greenwich Mean Time.) This clock is actually the eighth in a perpetually upgraded dynasty. Down the hall, behind a door that has been painted to look like a never-ending series of doors, each with a clock face displaying a different time, lie the bones of the ninth. These clocks' official names are Fountain 1 and Fountain 2, after their design type, the fountain clock.

Atomic clocks are extremely sensitive to environmental effects, so very few people get to see one operating. Burrus has worked around this by training a camera on Fountain 1, and he punches up a video feed on a giant screen in the hallway. Fountain clocks do not look at all like traditional timepieces: no faces, no pendulum, no moving hands. This one looks more or less like an upright metal tube, and most of the clock's action happens within it. At the tube's bottom, cesium atoms are heated until they gasify, and then launched into a vacuum chamber, where six criss-crossing lasers push them into a marble-sized supercooled ball, almost reaching absolute zero. The ball is tossed up into the vacuum tube and then begins to fall back down—hence the "fountain" name. On the way down, the atoms pass through a microwave cavity, which irradiates them at a certain frequency, causing them to have a spin state—either up or down.

The machine does this repeatedly, changing the tuning of the microwave cavity, until just about all the atoms are in the same spin state, meaning it's now tuned to the resonant frequency of cesium—exactly 9,192,631,770 cycles a second. Once you hit that mark, "you can count that 9 billion cycles per second and that's the duration of the second," says Burrus. "Count to 9,192,631,770. Boom! That's a second. And then you do it over and over and over and over and over and over."

The first operating atomic clock was built by Harold Lyons in 1949 using ammonia molecules, but the same principles. It was about as accurate as quartz crystal clocks, the era's standard, but atomic clocks kept getting more precise. In the 1950s, says Lowe, people switched over to cesium atoms, which are resistant to perturbation by temperature, humidity, and magnetism—all good qualities in a stable clock. Cesium became the international standard in 1967, and people began to make smaller commercial versions, which are now onboard satellites, airplanes, cell towers, even computer chips. (To be clear, these are atomic clocks, but they're not the government standard.) Today's accuracy level is like a clock neither gaining nor losing a second every 300 million years.

Atomic clocks aren't meant to measure the time of day, but interval time— in this case, the length of a second. "It turns out, in our very technological world, that's the more important commodity," says Lowe. That's because everything from wall sockets to cell phones to the world's GPS system runs off of frequency, or the number of cycles per second. "So the more accurately we can define the second," he says, "the more stable we can realize frequency." That makes a stable national power grid possible, enables cell towers to hand off calls to one another without dropping them, makes radar and GPS tracking more accurate, and allows a single twisted pair of telephone wires coming into the wall to carry not just voice, but high-speed Internet. All of that's more practical than knowing the time of day down to parts in ten to the 16th power.

But the atomic clock does make time-telling more reliable, too. Up until 1967, time had been defined astronomically, based on Earth's movement around the sun. "You would take the length of year, divide that by days, divide that by hours, divide that by minutes and you'd get a second," Lowe says. "But obviously that was very fungible because everything is in this cosmic dance." The Earth's spin and rate of travel around the sun vary during the year. Because the planet orbits the sun elliptically, it speeds up when it's closer and slows down when it's farther, so a day in November is longer than one in February. The Earth wobbles on its axis, a phenomenon called the precession of the equinoxes, and indeed the entire solar system wobbles. There's a sort of tidal effect within the solar system, in which Saturn and Jupiter exert a gravitational pull on the Earth when they are all on the same side of the sun.

Even the snowpack determines how fast Earth spins. "The majority of the landmass on the Earth is in the Northern Hemisphere, so if you have an exceptionally high snowy winter, that accumulates mass at a farther-out point, just like a skater moving their arms out slows the whole thing down," says Lowe, holding his out at shoulder height. "As all that snow melts and goes back to sea level, like a skater pulling their arms in, they speed back up." He crosses his arms over his chest and neatly pirouettes. And of course, over time, all systems lose momentum—everything slows.

Thanks to these variables, atomic time is a million times more accurate than rotational time, or measuring by astronomic movements. But rather than completely sever from nature, the atomic clock is actually an attempt to bridge astronomic and man-made time. Theoretically, if scientists counted by cesium alone, it would be possible, over vast periods of time, to end up swapping noon for midnight, thanks to the planet's eccentricities. None of us would notice within our lifetimes, and future folks might not even care; since the advent of electricity, the dark-light cycles that used to govern daily life have become increasingly ignorable. But atomic time is still synced to the sun. Just as we have leap years, we have leap seconds. They're erratically spaced, not occurring on any regular cycle. There were 36 of them between 1967 and 2015.

And atomic time is deeply affected by—and has offered insights into—the fundamental physics of the universe. When the United States was on its third atomic clock, Lowe says, scientists realized it would get different readings in Boulder than it would in Pueblo, about 150 miles away. They figured out that this was due to variations in the Earth's magnetic field, so future clocks were built with a magnetic shield. With the fourth clock, they realized they were seeing effects from stark shifts, or light-induced shifts in atomic structure, so they realized they needed to shroud the clock to keep it totally dark. With Fountain 1, Burrus adds, they realized black-body radiation was affecting the movement of the atoms, so the clocks are now cooled with liquid nitrogen. Each time you try to read time on a finer scale, you run up against new physical properties that affect your ability to measure it. The deeper you get into time, the deeper you get into nature. "Each generation of atomic clock has not only immediately led to applications that changed the way we live, but pushed our understanding of the physics of the universe, and indeed what is measurable," says Lowe.

Fountain 2, the clock that during my visit was waiting at the end of the hallway for launch, went online later in 2014. But it already has its own successor in the works. "We've reached a limit on what we can do with cesium," says Lowe. The next generation will be an entirely new design called an ytterbium optical lattice clock, or an optical clock, named because it uses visible light frequencies, not microwaves. "Parts in ten to the 18!" exclaims Lowe over their expected accuracy, bursting into a jubilant grin. "There's nothing mankind does that has this kind of level of precision. Not even close! We can't measure weight to parts of ten to the eight. We measure light intensity to parts in ten to the six." He sighs in admiration. "*Just incredible.*"

Physicist Andrew Ludlow and his colleagues have been working on the optical clock for over a decade now. And it looks even less like a clock than its predecessors do—more like a cluttered tabletop littered with dozens of tiny laser prisms and mirrors, a sort of gumball rally for light. While the optical clock uses a vacuum chamber and supercooling lasers, as the fountain clocks do, they're almost lost in its enormous jumble of parts. "Almost a Rube Goldberg device," says Burrus. "With photons!"

Instead of using a microwave cavity, this clock uses a lattice of laser light operating at what the researchers call the "magic wavelength" to trap the ytterbium atoms. Then they match the oscillation rate of a laser to the ticking rate of the atoms—about a quadrillion times a second, much faster than cesium. The higher the ticking rate, the finer the intervals into which you can divide time. It's like a ruler, Ludlow says: "If you get a ruler that has lots of little ticks on it, then you can measure very precisely." And with this higher performance, scientists can explore some really funky fundamentals of physics, including Einstein's theories about space-time. "One of the basic predictions of general relativity that Einstein made was that clocks tick slower in gravitational fields," a process called gravitational redshift, says Ludlow. Theoretically, a clock at high altitude should tick faster than one at sea level, because it is farther from Earth's center of gravity. Because this clock is fine enough to measure that effect, Ludlow says, "by carefully measuring, we can basically discern whether or not the original prediction by Einstein is exactly correct, or whether it's just an approximation to a more sophisticated theory."

Taking on general relativity is a pretty heavy task for one little clock, a poignant reminder of how much about the fundamental nature of the universe we still don't understand. The study of time forces you to think not only from the distant past to the vast future, but from the scale of the universe itself down to the very smallest things in it. "In a kind of philosophical way, time is maybe the least understood dimension of our reality," says Ludlow, gazing thoughtfully at the intricate machinery before him. "And yet ironically, from a scientific standpoint, it is the most well understood, because we can measure it better than any other physical quantity."

DEAN BUONOMANO'S LAB IS TRYING TO MEASURE TIME, too, at the level of the cellular network. The two-tone experiment he ran me through on my first visit was a psychophysics study, essentially, an animal behavior one. But he also wants to look at neural behavior, which is why Anubhuti Goel, a postdoctoral fellow in his lab, is seeing if she can teach rat brain slices to tell time. Or more specifically, she says, "We want to probe them for their ability to learn a particular pattern."

If Buonomano's intrinsic model of timing is correct—in other words, if the brain has no central clock—it should work on cells in a dish, dissociated from the larger structures of the brain. So Goel, a neuroscience researcher with thick black hair and a cheerful smile, has just returned to her electrophysiology rig from the lab's incubator holding a tray with six wells, each containing a shallow layer of tea-colored fluid. This is a nutrient medium, meant to maintain cells outside the body. Floating inside, on a piece of filter membrane, is a slice of rat brain, taken from the auditory cortex, its cells still alive. The slice is a ghostly off-white, incredibly thin, about the size and crescent-moon shape of a fingernail clipping.

Using tweezers, Goel picks up the filter and its slice and deposits them in a tiny clear dish. She places the dish into a silicone block called a chamber, slides the whole apparatus under her high-powered microscope, and loads in a series of tubes and wires that will keep the cells alive as she works. One pipes in artificial cerebrospinal fluid, which mimics the environment of the brain. Another brings in oxygen. Two heater arms maintain the bath at body temperature. The setup makes a constant low bubbling noise, like the murmuring of a fish tank.

As the cells get acclimated in the dish, Goel makes an electrode, inserting a glass capillary, hollow and no thicker than a strand of spaghetti, into a machine called a micropipette puller. A red glow flares as it melts the tube into two pieces with a sharp snap, producing a hollow pointed tip. She takes the finished electrode out of the puller, uses a syringe to fill it with a clear fluid mimicking that found inside brain cells, deftly flicks it a few times to get the air bubbles out, walks back to her rig, and is ready to begin.

Goel will "train" the network of cells by stimulating them in a pattern: one pulse, then a pause, then one more. The training can last from two hours to overnight. Then she'll see if the cells adapted to the pattern. This approach is sometimes called "learning in a dish," although Goel points out that the word "learning" is used broadly here, since it doesn't actually lead to animal behavior. For the stimulus, you can shine an LED at cells that have been genetically altered to respond to certain light wavelengths. (That's optogenetics again.) But today Goel will use short bursts of electricity.

She places the electrode in a manipulator that holds it at an angle over the dish, and then flips on a monitor showing a magnified image of the cells under the microscope. She zooms in with her lens, and the effect is something like watching a camera dive through a bowl of glass marbles. The cells are amorphously jellylike, clear with dark nuclei at the center. Goel finds one that she likes and presses the electrode up against it, making a tiny hole, allowing its internal solution to contact the liquid in the electrode. Now she can monitor the cell's electrical activity.

Goel turns toward a panel of electronics, including an old-fashioned oscilloscope, which reads out the cell's activity as green jagged spikes. "I am giving it a small pulse of electric current, which is about 50 milliseconds long. That is basically causing the cell to fire or release an action potential," she says, as a spike jumps on-screen. She tests the cell a few times, to see if it spikes more when she administers more current. It does. "This one is pretty good," she says, satisfied.

But, she points out, cells don't only respond to direct stimulation. Most of the time, they respond to a real-world stimulus, like a touch or a noise, which is being processed by an entire community of cells. So she also has wires running into the dish of cells, which will deliver current to the entire slice, including the cell she is monitoring. She touches a button, the current zaps in, and now on the oscilloscope, instead of one big spike, the cell makes several small ones close together. That's because the cell and its neighbors are all stimulating one another in reaction to the current. This is Buonomano's idea of throwing a pebble into a pond: The electric current is the pebble, and the activity in the neural community is the ripple it generates.

Now Goel is ready to train the slice by repeatedly delivering current to the dish: one zap, then a 100-millisecond pause, then one more. She wants to see if the circuit of nearby cells will learn the pattern. After she has trained them, she'll give a single zap and look at the cell's response. She points at the screen, which is showing a large spike for the cell's initial response. If she sees a second blip 100 milliseconds after the cell's first response, "then that means that the slice has learned this pattern," she says, as if the cell is anticipating the arrival of the second zap. The second blip, she says, is the response of the cell's neighbors, and implies the learning is a coordinated effort among them.

This is important, because Buonomano's model hinges on the idea that information about time is encoded in the changes in activity among collaborating neurons, or the pond's ripples. If their lab can teach a circuit to reproduce a certain pattern, Goel says, "it would support the theory that yes, it is possible for a circuit to actually give you information about time just based on the change in activity patterns of the cells in the network." And because these are just cells in a dish, not the entire brain of a living animal or even a specialized region, she says, "that would rule out the requirement of some special organ that is doing some sort of emission, or some sort of counting or integration."

The work is ongoing; they probably won't have more than preliminary results for a few more years. Until then, Goel's job is to keep throwing stones into the pond, measuring the ripples.

NOW THAT WE'VE SEEN TIME on the very smallest scales, let's go back to London, where Rooney is patiently waiting at the prototype, an ode to the enormous, both in size and in lifespan. The clock's power—even as a still-under-construction object—is that it forces you to think about a future in which you no longer exist, to imagine time as infinitely bigger than yourself.

"It's a focal point for conversations," says Rooney, nodding at the crowd passing by. "In the context of this gallery, it might make you think about how things are made." And there are some very special considerations that go into making a device that will stand throughout civilization. The design had to be what Rooney calls "future proofed," inured to calamities that might confront a very long-lived object, one that might spend most of its life without a custodian, as he has been to the prototype. Inspired by the foundation's study of other long-term architectural sites—the Svalbard Global Seed Vault, nuclear waste facilities—it was sited in the high desert to protect it from water damage, and away from cities, with their vulnerabilities to wars, crowds, pollution, and vandalism. Most of the metalwork is made of marine-grade stainless steel, calculated to lose only a few thousandths of an inch to oxidation over 10,000 years; the rest is high-grade titanium. It functions without lubrication, which would need to be replenished, instead using ceramic bearings. The weight system will store power in case an extreme snowpack or the dimming of the skies after a volcano or an asteroid strike prevents sunlight from getting through for a while.

But it also has to be future proofed to ensure that people will understand what the clock is and how to take care of it. By then they might speak new languages, or be *less* technological than we are. The clock might break, or be lost and wind down before being rediscovered. For a future person to fix it, says Rooney, it must be "understandable by inspection"—the parts' functions visible to the naked eye, without microscopes or diagnostic tools; fixable without having to take the whole machine apart; and requiring no more than Bronze Age technology to repair.

"The idea the foundation had was to have the mechanical computer, in the same vein as the nineteenth-century Charles Babbage's mechanical computers," Rooney says, gesturing across the room to, in fact, that very thing. The museum has the prototype of Babbage's Difference Engine, the granddaddy of all computers, to which the clock bears a striking resemblance. Both count through the motion of rotating columns of geared wheels, although the clock is binary and the Difference Engine is decimal based. Both look strikingly futuristic and primitive at the same time, computers without screens or lights, their insides naked to the world, metal on metal.

Compare that to the object sitting between them: the first Cray supercomputer. It's a tall C-shaped column, with beach ball–esque blue and red stripes running up the sides, and it houses racks of printed circuit boards. In

1976, the Cray-1A was the world's fastest computer, with a 160-megaflop speed and an 8-megabyte memory. Only four decades later, it seems profoundly dated and technologically opaque. Forget future civilizations; if most of us were to stumble across one buried in a remote cavern, we'd be hard pressed to identify it as a computer, much less fix it without a manual, an engineering background, and some very specific parts. (It contains some 50 miles of wiring and requires Freon to cool.)

This is why the foundation decided to build a mechanical device, instead of an electronic one or a virtual online ticker, says Rose a few weeks later in San Francisco, when I stop by to see how things are coming along. He's taking a lunch break across the street from the Interval, the hybrid bar, café, library, and office quarters the foundation was then about to open to offer a salon for lectures on future-oriented topics and to showcase some of the clock's mechanical prototypes. "If you arrive at a thing and there is an LCD screen and it's blank, you can't tear into the silicon of it and understand what the intent was," says Rose. And with electronic objects, he adds, "we kind of consider them magic, so we're not impressed by them anymore. They can do *anything*. So making the thing mechanical, where you can see how that's all working, is much more interesting. Half of the goal is to make it work. The other half is to make it interesting."

After all, you've got to make the hike through the desert worthwhile, and encourage people to preserve what they find. The clock itself is a giant Easter egg hidden in a mountain, and within it will be more surprises: the unique chime every day and "anniversary chambers" that will display special objects on certain milestone dates.

They've planned for the site to be a tourist mecca or wholly abandoned, for the clock as a new object and for its life as a ruin. "We designed it to be broken," says Rose, inspired by objects like the Antikythera mechanism, a first-century bronze-geared "computer" used to track astronomic positions, discovered underwater off the Greek coast in 1901. While many mysteries surround the device, including who made it, scientists now have a basic understanding of how it worked and how to build one—there are everything from virtual to Lego reconstructions of it. "The Antikythera device was broken for 2,000 years," says Rose, but "it still works in the sense that we understand what it is."

But making something work is not the same as understanding what it means—and here it might be useful to be enigmatic. Rose points out that many artifacts from the deep past are more interesting because we *don't* quite get what they were for. Take Stonehenge. If its mission statement "had been etched into each stone," he says, "we'd probably be like, 'Oh wow, this was this really silly druid cult who sacrificed virgins.' The answer would have been less interesting than the mystery."

So the clock design works to inculcate mystery, starting with the pilgrimage built into the viewing process—the remoteness, the difficulty, the required ascent from darkness to light. The Long Now studied historical and land art sites—*Star Axis*, the Trinity bomb-testing site, the pyramids, Petra—to compare experiences visitors could have. One of the most influential, Rose says, was visiting *The Lightning Field* by Walter De Maria, a grid of stainless steel poles in the New Mexican desert that you can only view by staying overnight in a small cabin, with no guarantee that you will see lightning. "It kind of tricks you into spending 24 hours completely isolated in a beautiful desert landscape, which is in and of itself a powerful experience," says Rose. That period of contemplation, of wonder, of humility before nature, of making a journey with friends and discussing what it all means—that is where he thinks the magic lies.

No doubt, people are going to come up with their own ideas about what the clock means. And that's OK with the foundation, although Rose points out that they have been careful not to align their project with religious or national ideologies. So far none have lasted for 10,000 years, and it's not clear whether they're ultimately protective or destructive. People who support your ideas might work to preserve your artifact, but if the idea loses dominance, it could be abandoned or even targeted for attack. Rose likes to use the example of the Buddhas of Bamiyan in Afghanistan: fairly innocuous for religious symbols, he points out, and seemingly future proofed. They were remote, in the high desert, and difficult to destroy. "Yet the Taliban spent a lot of effort, ordnance, explosives blasting these giant beautiful buddhas out of the wall," says Rose. "And it was basically because of the myth that went along with them—it didn't agree with the myth of the people that were encountering it." So the foundation is trying to drive conversation about time more than any particular narrative. You just have to make something "mythic in proportion, both in time and in physicality," says Rose. "And then hopefully other people will write that myth."

He pushes back his chair and we walk across the street to the new café, where he has one last task to do before it opens to the public. The library shelves are only partially filled in, but already they're heavy with time: Stephen Hawking's *A Brief History of Time*, John McPhee's *Annals of the Former World*, and *Swann's Way*, the volume of *In Search of Lost Time* in which Proust encounters the madeleine. The bright scent of lemon wafts across the room from where the bartenders are zesting citrus for cordials; a delivery guy keeps rolling in giant boxes of Chinotto and tomato juice. Bar manager Jennifer Colliau is chatting with the staff as they finish learning how to make the time-themed drinks she's devised for the menu: an aged tequila-and-pomegranate-juice drink called ponche de Granada, a curated history of the evolution of the martini, re-creations of the daiquiris served at La Floridita, the Cuban bar that was Hemingway's hideout—a nod to both literary history and the Prohibition era.

While the clock is still being fabricated elsewhere, pieces of it are here. The bar top is made from stone slices carved from the mountain bore. The first version of the chime generator holds up a glass-topped table. And at the entrance stands a model of the orrery, a gyroscopic model of the planets you can see from Earth using the naked eye, each one orbiting the sun on a metal ring. It's mounted atop the same binary mechanical adder used by the London museum clock prototype, which will swing the planets through their timed orbits. There will one day be an orrery inside the mountain, too, tracking the position of the planets, perhaps crowning the clock or sealed off in an anniversary chamber, a surprise for lucky visitors.

These planets will be a final point of reference for a future people who may have outlasted all of our calendars, who may understand the clock only by matching its parts with what they see in the heavens. "Since we've had electrification, we've largely ignored the night sky," says Rose. But before the last century, "the night sky played an extremely big role in civilization. It is fundamentally the clock by which all of humanity has ever timed themselves. And so it's the thing that we wanted to tie the clock to."

From a battered cardboard box, he begins pulling out spheres encased in bubble wrap and painter's tape. Each is a stone representing a heavenly object. He teases the wrapping from the sun, and holds it aloft, a softly glowing ball of yellow calcite, then slips it onto a peg in the orrery's center, as delicately as setting a star atop a Christmas tree.

"Mercury first," he says, unwrapping a pewter-colored ball, actually a bit of meteorite, and using a tiny pin to fit it onto one of the rings.

Next is Venus, a luminous peach calcite. He carefully fixes it to the next ring.

"Earth is a Chilean lapis," he says, holding out a blue and white marbled orb. "It's like a cloudy lapis that has some granite and quartz in it." And then he slips it onto a pin and anchors it into the orbit of the sun, a message from our present to our future, a tiny marker in both space and time: We were here.

SEVEN

· ·

Pain

THE NIGHT LIGHT is all dark wood and leather and red and gold flocked wall-paper; the bar's time era seems set somewhere between "Gold Rush" and "impossible future." The cash register is a vintage machine. There's a semispooky antique wooden figurehead floating over the bar. Over the course of the evening, the bright piles of citrus fruit piled into bowls along the bar will be cranked through an extremely steampunk hand press. Behind the bar is John Nackley, bearded, wavy-haired, wiry, clad in black jeans, checked Vans sneakers and a t-shirt advertising a local skate shop. He's the star of the show, cheerily greeting customers, mixing cocktails, making change. He and his co-owner designed the Night Light as a neighborhood lounge where everybody would feel welcome, a retro-future hideout "like if there was a bordello aboard Captain Nemo's *Nautilus*," he says. He pauses thoughtfully. "Do you remember his room with the big pipe organ? You know, it was very ornate. Really *lush* for a submarine."

It's Happy Hour, and as the Fresh Jamz deejay crew spins reggae and dub in the background, the after-work crowd at this downtown Oakland watering hole seems pretty, well, happy. But not me. I am here because I want to know about the nature of pain, specifically the pain of social rejection. And who has observed that very bitter form of pain more often than your local bartender? Nackley's been pouring drinks up and down the West Coast for 16 years. That's given him plenty of opportunities to hear tales of romantic woe and to dispense the occasional bit of what he insists is bad advice. "I'm no psychologist, and I'm no smarty," he says leaning over the bar, "but I'll tell you what, I can tell you a thing or six about broken hearts."

And broken hearts are why I'm here. One of the most fascinating wings of pain research has to do with the question of how the brain processes physical pain, or the pain of a broken bone, versus social pain, or the pain of a broken heart. Led by researchers like Dr. Naomi Eisenberger, a social psychologist at

UC Los Angeles, some researchers argue that these processes are astonishingly similar. As with the taste researchers, one of their clues is language, because so many people use physical terms like feeling crushed, bruised, or heartsick to describe rejection. But taste and pain researchers are essentially contemplating inverse arguments: While the food science folks are debating whether we could perceive a new taste if we established a discrete word and conceptual category for it, pain scientists are wondering whether two experiences we perceive as categorically separate are in fact neurally the same. In other words, instead of thinking about splitting off new perceptual categories, they are wondering if we should *collapse* them. "It seems to be a universal phenomenon that people describe these negative experiences associated with breaking social bonds as being hurtful," Eisenberger says. "This is sort of intriguing, and led us to wonder whether there is something that is actually really painful about social rejection or social loss, or whether it's just a figure of speech."

Armed with a decade of studies, most of them done by putting people in fMRI scanners, they now believe it's no linguistic coincidence, and that the brain interprets social pain in a way that is similar to—*and just as real as*—physical pain. The pain of heartbreak may be "all in your head," but now they're pretty sure exactly where: the dorsal anterior cingulate cortex (dACC) and the anterior insula (AI). These areas process physical pain, but in study after study also activate during social rejection. Social rejection work opens up some really intriguing questions about what pain is and how we perceive it, questions that have been unscientifically picked apart in the friendly gloom of places like the Night Light for years.

Now, you may be wondering, why hang around bars to see how Eisenberger's ideas play out in ordinary life, instead of, say, sitting in with licensed mental health professionals? Two answers: First, bars are not bound by the patient confidentiality rules designed to protect sensitive interactions between therapist and client. Second, bars, and perhaps hair salons, are on the short list of public places where people feel emotionally intimate with strangers. They are free spaces for talking honestly and in ordinary language about bad feelings, where even the reticent might open up about what's bothering them, especially if they've got a friendly captive audience behind the bar.

And Nackley is happy to listen. Ladies usually ask him to interpret men's behavior. "Women will say, 'What does it mean when a guy does this and that?'" he says. Men are different: "With the fellas it's almost always 'She cheated.'" Nackley has heard plenty of tales of betrayal. There was the guy whose wife wasn't taking business trips. There was the woman who found out that her husband had a girlfriend on the side. Every now and then, people will come in with a nonromantic rejection; they didn't make the team; they lost the job. But mostly, they talk about love. People don't plan to talk to him, Nackley says. The combination of a solo trip to the bar and a little booze means that things just spill. And that's fine with him. "I am qualified to do maybe three things," he says. "Ride a skateboard, and make drinks, and the third one just happens to be listening."

In his years tending bar, Nackley has put his finger right on a few of the broad themes that labs like Eisenberger's are exploring through clinical methods. Yes, he says, people use the language of physical pain to describe social pain, and people going through breakups come in complaining of physical symptoms: stomachaches, lack of sleep, even graying hair. But more than that, he says, his own experience and that of his customers lead him to believe that social pain is more bothersome. "Just as an example, I've got like a heel issue right now. And it hurts. Standing here is physically painful for me," he says, leaning against the bar. "But I suffered far more a few months back when my sweetie of three years gave me some bad news."

Social pain also lingers longer. "Physical pain is oftentimes much shorter lived," Nackley says. "You know, you get over it, that broken leg or torn ligament or something like that. You get over the physical thing. You can take Advil."

And, speaking of Advil, you can dampen social pain with drugs. By trade, Nackley dispenses a legal painkiller: alcohol. But even a drink and a talk, he says, are "only a short-term remedy. It's like taking a Tylenol, you know. It doesn't cure the problem at all. All it does is mute, or turn the volume down at least, on the symptom that's causing you the aggravation."

And it's interesting that he mentioned Tylenol in connection with heartache. Because in the academic world, that is where this new wave of pain research has taken off.

NAOMI EISENBERGER'S OFFICE overlooks the sprawling UCLA campus. She's been here her entire career, starting as a graduate student in health psychology. She was intrigued right off the bat by the connection between the social and the physical—"How is it that what goes on in our heads seems to influence what goes on in our bodies? Why does stress make us sick?"—and drawn to the neuroscientific techniques that have made these connections increasingly possible to examine.

She got hooked on studying social pain from the very beginning. "I think I have just always been curious about rejection," she says in a soft, soothing voice. "Why does it seem to affect people so much? A lot of people have memories of early childhood experiences of being picked last for teams or left out by their friends on the playground." In her own life as a grad student, she'd noticed this fear of rejection showing up as nervousness about public speaking. One time, when she had a quiet moment by herself before a speech, she became suddenly aware of how rapidly her heart was beating. "It really feels like I'm being held up at gunpoint," she thought to herself, "and this is weird, because all I'm doing is giving a talk."

Eisenberger began studying the brain activity of people who had been socially rejected as part of a lab experiment. One day as she was looking at her data, she happened to be sitting next to a friend who was analyzing data from a pain study of patients with irritable bowel syndrome. "We just sort of noticed, 'Isn't that weird? The activations that you are seeing in your irritable bowel syndrome patients who are being exposed to painful stimulation look really similar to what we are seeing in this rejection study,'" she recalls. "These two things, maybe they are more similar than we thought. Maybe it's *not* just a metaphor."

Now if you want to get to the bottom of whether social rejection actually hurts, the first dumb question you have to ask is, well, what is pain? And it turns out that the answer is not so obvious. When I ask Eisenberger, there's a long pause. "That's a super hard question!" she finally says with a light laugh. "And I think depending on who you are talking to, different people care about different aspects of pain."

For the record, she points out, there is an official definition, issued in 1979 by the International Association for the Study of Pain, a group of scientists, doctors, and others who research and advocate for pain relief. Their definition is "an unpleasant sensory and emotional experience associated with actual or potential tissue damage, or described in terms of such damage." That's incredibly broad; it really tells you a lot more about how pain feels (bad) than how it works. But it's telling that it encompasses the very linguistic mystery that Eisenberger and her colleagues set out to unpack. What is a broken heart if not an emotional experience described in terms of tissue damage?

There are reasons why describing pain is so hard. For one thing, it's difficult to objectively measure something that is inherently subjective, points out Dr. Sean Mackey, chief of the Division of Pain Medicine at Stanford University, whose lab has also researched the idea of overlap between social and physical pain. How do you turn the sensation of pain into something you can *count*? "There is not a direct one-to-one correspondence between a specific quantum of stimulus and experience of pain," Mackey says. How much pain a person experiences from a given stimulus can vary greatly—what is awful for one person might be tolerable, or even barely noticeable, for the next. Without an objective way to measure how much pain a person is in, medical and mental health practitioners must rely on the same feedback mechanism: the patient's self-report.

Pain is also polysensory; we feel it through many channels. People often think of touch first when it comes to pain, and some researchers indeed classify pain as a subset of somatosensation, the larger category that includes touch and temperature. We have nociceptors, or pain sensors, throughout our skin and soft tissue that are sensitive to environmental changes that might cause us bodily damage—pressure, temperature, chemical acidity. These nociceptors let us know when we've pinched our fingers in a drawer or burned our tongues on hot pizza or gotten shampoo in our eyes. It's important to note that when we experience pain this way, it's not because we've overstimulated the regular touch mechanoreceptors. We've actually activated an entirely separate system of receptors that don't kick on until the force, temperature, or chemical irritant we are experiencing reaches a certain dangerous level. These impulses are relayed to the brain through a pathway separate from touch.

But, Mackey argues, you can experience pain through any of your senses, not just touch. Ordinary light doesn't hurt the eyes, but if the light's too bright, he asks, "doesn't the light stimulus then become painful? And the same with sound. If you happen to have your ear next to a gunshot, isn't that painful? You are exceeding a certain threshold for the sound pressure waves to be perceived as painful. What we believe is that these other sensory inputs can actually engage the same type of pain systems as if you hit your thumb with a hammer."

That's an important idea: Pain has multiple sensory pathways that all feed back to the brain. Technically, Mackey says, what happens in the body (what a neuroscientist would refer to as the periphery, made up of the nerves and the spinal cord) is not exactly pain. It's nociception, or the translation of real-world data into electrochemical signals signaling pain. Those signals get piped to the brain, where perception truly happens. "Pain is fundamentally a brain-related phenomenon," Mackey says. The brain is where it all registers, "where the perception of pain is processed and perceived and modulated."

Another complication is that pain has several components, although not all researchers tally them up the same way. Eisenberger likes to speak of pain as having two main parts. The first is its sensory component, which is mainly objective information: Where is the pain coming from on the body, how intense is it, what is its nature? For example, she says, "is it a burning pain or an aching pain?" The second is its affective or emotional valence, how distressing or bothersome it is, and your urge to reduce its unpleasantness. Mackey thinks there are at least three components, possibly four. The third he calls the "cognitive evaluative" component, or your thought processes about how to get away from the pain and what the pain means. The fourth, which he says is less accepted and perhaps related to the third, is the idea of behavioral avoidance, or doing things to prevent future pain. In fact, that behavioral and motivational aspect of pain is probably the key missing component of the definition of pain, Mackey says. (Some experts combine these last three categories under a broader "affective-motivational" heading.)

Different brain areas seem to be in charge of handling these dimensions of pain. As you might expect, the somatosensory cortex, which is involved with sensing touch, is involved with sensory pain. The anterior cingulate cortex and insular cortex—involved in processing emotion—are involved with pain's affective dimension. The prefrontal area, which is involved in planning and decision making, is linked with its cognitive aspects. But, says Mackey, there's really no clean break between these areas, which function as part of a larger system. "All of these regions are intimately connected to each other and each one is modulating the others," he says. Many researchers refer to this as the "pain matrix," says Eisenberger, a distributed network of regions that activate when you feel pain. "Some are involved more in sensory components, and some are more involved in the affective experience," she says.

And it's here, within this idea of overlap and blur, that we get to Tylenol and lost love and fMRI scanners. If these areas are truly cross-chatting, painkillers that work to calm muscle tension should work to quell heartache, and vice versa—love should be a balm. Or in experimental terms, says Eisenberger, "if we turn up physical pain, does that turn up social pain? If we turn down social pain, does that turn down physical pain?"

This idea has its roots in the 1970s, when neuroscientist Jaak Panksepp realized that giving infant monkeys morphine—a potent painkiller—made them produce fewer distress cries when separated from their mothers. It was an important clue that an analgesic for physical pain reduced social pain. Other research avenues have explored how psychological factors can influence physical pain perception, like how the context of pain changes how strongly you feel it. Then there's the placebo effect: Why do people taking inactive pills report that they feel better? But Eisenberger's group was the first to test Panksepp's idea in humans by putting people into a scanner and, well, rejecting them.

It's actually hard to reject someone who is lying inside a giant magnet. You can't get anyone else in there. They're not allowed to talk or move. It's so noisy that they can't really hear. But they can play Cyberball. Cyberball is the brainchild of Kipling Williams, a psychology professor at Purdue University, who came up with the idea after being slowly excluded from a real-life game of Frisbee that he'd run across in a park. In Cyberball, study subjects are asked to pass a virtual ball back and forth with several other players. At first, the other players pass the ball back. Then they start ignoring the subject, making it a game of virtual keep-away. The other "players" are actually a computer, programmed to eventually exclude the person. But the subject doesn't know that, and feels stung by the snub.

In their first 2003 study, Eisenberger and Williams' group found that rejecting Cyberball players caused greater activity in the dACC and AI, both regions otherwise associated with physical pain. And over the next several years, Eisenberger's lab explored variations on this theme. They found that people who score high on tests for sensitivity to rejection have a heightened dACC response when shown images of disapproving faces. People asked to participate in an interview and then get feedback from an "evaluator" (really, a lab researcher) while lying in the scanner showed a bounce in dACC and AI activity after hearing themselves described with words like "boring" that connote rejection, but not after hearing neutral or accepting words. Teenagers who spend more time with friends show *less* activity in these pain areas when rejected during Cyberball.

Other labs were exploring, too. One particularly interesting 2011 study, led by social psychologist Ethan Kross at the University of Michigan, asked people who had just been through unwanted breakups to look at pictures of their exes, arguing that this painful stimulus would be even more acute than being left out of an imaginary game or criticized by strangers. Subjects lying in the scanner either looked at a picture of their former partner and thought about being rejected by them or viewed a photo of a friend and recalled a recent positive experience with them. To establish a baseline of which brain areas react to physical pain, a separate group of subjects was scanned while feeling either painfully hot or neutrally warm stimulation on their forearms. (Pain in these experiments is typically administered to the arm using a small wand with an electric thermode at the end that delivers a sharp heat; it feels, Eisenberger says, more like a sting than a burn.) The researchers found that not only did people report more pain when looking at their exes, but their brains showed more activity in the dACC and AI areas—the same ones that became more active for the people touching the hot object.

With the evidence mounting that social pain inflames the brain's physical pain centers, it was time to try the reverse: to see if you could use physical pain remedies to calm social pain down. In 2010, social psychologist Dr. C. Nathan DeWall at the University of Kentucky, collaborating with Eisenberger and others, tested the social pain-killing power of Tylenol, or rather, the generic acetaminophen. DeWall first asked his subjects to take either acetaminophen or placebo pills daily. Every night, they logged how much social pain they had experienced that day using a "Hurt Feelings Scale" developed to gauge the pain of rejection, but not other negative emotions. They also recorded their day using a separate scale that measured positive feelings. After three weeks, the subjects taking the acetaminophen reported fewer hurt feelings than those on the placebo, but not an increase in good ones, suggesting that the drug was tamping down bad feelings, not enhancing the positive ones.

In the next stage of the study, DeWall's subjects once again took either acetaminophen or a placebo for three weeks, and then got in the scanner to play Cyberball and be roundly rejected. The participants who took the acetaminophen showed less activation in both the dACC and the bilateral anterior insula. (Interestingly, while their brain activity differed, being left out of Cyberball *felt* equally distressing to both groups.) These results, DeWall says, suggest that "we put all of these different painful or unpleasant events in separate buckets in our heads, but there is a common mechanism underlying them."

So should doctors start prescribing Tylenol for people going through breakups? "I don't know," DeWall muses. While the authors didn't go so far as to recommend that people start routinely popping Tylenol to inure themselves to negative feelings, they did write that it might offer temporary relief from social pain, and suggested further research to see if it can also dampen the aggression and antisocial behavior that can follow rejection. Since the study came out, DeWall says, he's gotten a lot of letters from people sharing anecdotes about their own attempts to self-medicate for a broken heart, but so far there's been no clinical trial testing Tylenol on the lovelorn.

There's an X factor, too, in that it's not very well understood how acetaminophen kills pain in the first place. "Does it work on central pain versus peripheral pain?" asks DeWall. "Honestly, we don't know enough to make a definitive statement about it." But he does know that it activates cannabinoid 1 brain receptors, which are also activated by THC, the psychoactive component of marijuana. In 2013, along with several collaborators, he published the results of four studies investigating the effect of pot on social pain. The first three were correlational analyses, in which they argued that marijuana use correlates with lower self-reports of loneliness and incidents of serious depression, both indicators of social alienation. The fourth asked people to play Cyberball, but only half of them got a version in which other players excluded them. Afterward, the players filled out a scale that assessed how threatened they felt their emotional needs—self-esteem, belonging, control—were during the game. Frequent marijuana smokers reported feeling less threatened than the infrequent ones. Again, the authors didn't suggest everyone light up to avoid social pain—in fact, they wrote, people might smoke pot *because* they feel socially rejected. But they did suggest that both drugs suppress social pain by acting on the same cannabinoid 1 receptors, and pointed out that once again a drug that is—at least in some states—legally used for physical pain seems to also alleviate social distress.

JOHN NACKLEY IS PREPPING THE BAR, getting ready for another Friday night, decanting olives and maraschino cherries into smaller containers. Sitting across from him is Kerry, a woman with a cute bob and long dangly earrings. (We'll go first-name-only here as a privacy consideration for bar patrons kind enough to share their romantic woes with strangers.) Kerry is nursing a beer and an unrequited crush that won't die. This is a devilishly tough kind of romantic rejection to get over, because, as she says with a sigh, "this never really began."

Her story is an old one: Girl likes boy. Boy likes motorcycle. Boy meets a different girl and her motorcycle. "And that was it," Kerry says.

But it wasn't, really. Kerry still feels terrible. She'd hoped they might be more than friends. But even if that didn't work out, after a couple of years of hanging out together, she thought they'd at least be pals. "I really thought we were going to be friends til the day we died," she says wistfully. "I've never had such a close friend." But after the new girl came along, she and the guy had some words. Now they don't talk anymore.

So instead she talks to Nackley. The two of them had known each other from their hometown, but they'd fallen out of touch after high school. Kerry looked him up on Facebook just as her friendship was ending and the Night Light was opening for business. When things were at their worst, talking to Nackley became her emotional outlet. "I came in early, so there was no one else here, and I just, like, sobbed into my beer, literally," she says. He listened. He gave her friendly advice. He cued up songs on the bar's record player, each one designed to either soothe the wounded heart or wind it up even further. "Ever Fallen in Love?" by the Buzzcocks. "One Less Bell to Answer" by the Fifth Dimension. "Anything, Anything," Dramarama's ode to obsession, which pleads, "I'll give you candy, give you diamonds, give you pills, give you anything you want, hundred dollar bills."

Now, says Nackley, "It's a pretty regular therapy session we have going on here." Kerry drops in once a week. He brings her a beer. They chat. Tonight they have an interloper who wants to know if the pain of that romantic rejection is anything at all like physical pain. "I would have rather broken an arm or a leg," says Kerry drily.

"Hear, hear. I'm with you," says Nackley, now slicing limes.

Social pain, they agree, just hurts more. But why? They think it has to do with uncertainty, how you never know when it will end, how you can't see the healing process the way you do with a bodily wound. "I have faith that my bone would heal," Kerry says. "But I don't have faith that I will ever meet someone that I will actually have feelings for."

That uncertainty, they say, also underlies those unanswerable questions about self-worth that get raised when your loved one leaves you: Were you not good enough? What did you do wrong? Did you miss the signals that your partner was unhappy? What's so great about the new gal? "Nobody should be asking themselves those kinds of questions," Nackley says, going off to do something at the cash register. "Those are terrible questions."

(For the record, Nackley tells me another time, this is not at all the kind of obsession and self-doubt that he hears when people come into the bar upset from other forms of social loss, like the death of a loved one. That pain is missing the uncertainty and self-recrimination of a breakup, he says, "because when someone dies, that's absolutely permanent. There's no going back." With breakups, he adds, "I think people always hope in the back of their mind, 'Oh, maybe we can still get together if I can just change their mind.'")

So if you don't know exactly what went wrong or when you'll feel better, you stew. In Kerry's case, she says, "I was obsessing and I couldn't stop thinking about it. And I had to talk about it or I would have gone crazy, you know?"

The interloper wants to know if talking about it actually helps. Nackley returns, leans over the bar. "I think it makes it worse," he says in a dramatic whisper.

"You do?" Kerry asks, startled.

"Well, we keep bringing it up. She'll feel better about it for a minute, right?" But then, Nackley says, turning to her, "you're like 'Oh, man, I miss him.'" Nackley shakes his head and puts on an admonishing tone. "Stop it, Kerry! Don't miss him. Don't miss him. He's a turkey, I'm telling you."

"It confuses me," Kerry says glumly. "One minute, I'm missing him and remembering all the good things. And the next minute I'm angry at him." And it's this mental agitation, this unending ping-pong between hope and despair, the two of them agree, that makes social pain outlive physical pain.

"It's harder to get away from," Nackley says.

"How do you get away from *your mind*?" Kerry asks.

YOU MAY NOT BE SURPRISED that Naomi Eisenberger has tested this idea by putting people in an fMRI scanner. Specifically, she wanted to know if remembering a broken bone really does hurt less than recalling a broken heart—if we can relive social pain more readily than physical pain. "People are really good at being able to remember and relive social pain experiences," she says. "So we can sit in our office and think back to when we were dumped by our high school boyfriends and sort of get back into that emotional state and how awful it felt. We can't do the same thing with physical pain."

This is not a casual statement. Because she is so interested in the topic, Eisenberger has spent a lot of time trying to relive the most physically painful event of her life: childbirth. She doesn't mean she's tried to remember its emotional moments; she means its actual visceral qualities. "I remember it being *intense* pain," she says. "But when I try to sort of go back into the body and re-create those painful feelings, nothing happens."

When her lab started looking into reliving pain, she once again worked with Williams, who had previously been involved in a study showing that people could intensely relive social pain many years later simply by writing about it. In Eisenberger's lab, people were asked to write journal entries about both social and physical experiences that had caused them a great deal of pain, and to rate how painful those experiences had been. Then, as the subjects were lying in the scanner, they were cued with lines from their journals, and prompted to relive the experiences. The lab members were monitoring two things: First, they asked the subjects to rate how much pain they experienced as they recalled each memory. Second, they looked at which areas of their brains became more active.

The participants found it more painful to reexperience social memories than physical ones, even when recalling events that had previously been rated equally painful. And there were some differences in the brain areas that lit up in response. One of the most prominent areas for physical pain was the lateral surface of the brain, which is involved in sensory perception and monitoring the state of the body in general. For social pain, a more active area was the dorsomedial prefrontal cortex, which is involved in thinking about mental states and the minds and intentions of others.

While this alone might not explain everything, Eisenberger points out, it's an intriguing difference. When you are processing a social rejection, you are not thinking about your body, but about how *others* are thinking about you. "So maybe in reliving the social pain, people are sort of thinking about 'Why did that person dump me? What does that say about me? What were they thinking about me?'" Eisenberger says. And as you recall that breakup, you can reengage those affective regions in a way you can't reengage your sensory regions. You can repeat the stimulus of the social snub in a way you can't repeat the stimulus of a physical injury without a fresh cut or a newly stubbed toe.

DeWall points out that because social rejection can be so ambiguous, and because the minds of others are so mysterious to us, we can keep returning to that pain, puzzling it out much longer than we would with a physical injury, where the cause is obvious. "I cracked a vertebra in my neck playing football when I was in high school, and that was enormously physically painful, but I'm not thinking about it today. I'm not thinking, 'My gosh, what did I do to make that happen?' Well, I played football and sometimes people who play football get hurt," he says. "But if you ask me to relive the most painful social rejection in my life, suddenly it's a different playing field."

We ruminate over rejection because we want to understand it. And while there may be benefits to doing that, when it goes on for too long, it can become a problem. "One of the best and the worst parts of the human mind is our ability to try to make sense of things," DeWall says. "A lot of times it helps us. But occasionally I think it can sting us. And nowhere is that more evident than with the pain of rejection."

NOW, THIS IS NOT A SCIENTIFIC ASSAY of all bartenders and their patrons on the pain of lost love. But I do think that if you're going to spring a weird conversational topic on people who are just trying to have a drink, you should see if your results can be replicated. Maybe John Nackley is just a remarkably insightful observer of the human condition, or maybe Kerry just happened to be the one person in the world who would make the broken heart/broken bone connection on the first try. Or maybe there is something about the Night Light and its bordello vibe that invites the philosophical contemplation of romantic disaster, because all of the other folks I talked to on my visits there had basically the same things to say as Kerry.

So I'm at another bar, and it's the polar opposite of the dim retro cool of the Night Light. It's called Lefty O'Doul's, and it's a family-style hofbrau/sports-pub right off of San Francisco's ritzy Union Square. It is two weeks before Christmas, and the place is a madhouse. The line for food goes out the door. The interior is all shtick: Christmas lights and photographs of celebrities and baseball-themed bric-a-brac. In the very back of the place, Lisa Mongelli is serving drinks at a bar that is located next to a statue of Marilyn Monroe caught mid-skirt disaster, and a Christmas tree that is, for some reason, upside down.

Mongelli is an energetic, raspy-voiced pro-snowboarder-turned-bartender-turned-drummer. Her long dark hair is twisted back in a rubber band with a pen stuck through it; her rolled-up shirtsleeves expose a sleeve of tattoos down her right arm that combine a bass clef, flames, the California poppy, and the motto "Any vice worth having is expensive." She is, perhaps, the most enthusiastic bartender it is possible to meet. She started working at a club at age 18 while living in Australia (the drinking age is lower there) and immediately took to living in what she says is essentially a real-life version of *Cheers*. She loves the constant stream of new people, the conversation, the chance to make someone's night with some friendly attention. She calls everyone "sweetie" or "love." She is not faking this camaraderie. "I get paid to have a conversation. I get paid to entertain," she says earnestly. "The job is not to just make drinks."

And this is an exceedingly happy bar, not just because it's Christmas, and not just because it is located at the back of a jammed, noisy, brilliantly lit restaurant that is also selling enormous slabs of roast beef, but because Union Square is basically a gigantic tourist trap. Most of the people here are on vacation, coming in merrily from outdoors where they've been enjoying the square's sky-high holiday tree and the hot chocolate and the ice skating rink that somehow manages to exist even though it's about 70 degrees outside.

Mongelli says she loves working here because the bar is so crowded that on a good night she will make about 200 new friends. If she does her job well, they will do likewise. "Everybody here is awesome," she says, pointing out that 90 percent of them have traveled here specifically to have a good time. "Ask anyone where they are from," she offers. There's the guy to my right, who I think is German because he correctly pluralized Hefeweizen, but he turns out to be from Luxembourg. There are some shy Australians eavesdropping on our conversation just around the bar's curve. There's the couple from Nevada who stopped in from holiday shopping to order up a pair of Moscow Mules. There are Steve and Maureen, the middle-aged couple sitting to my left, who have driven in from elsewhere in California. And there are friends Alex and Russ, the only regulars here tonight. They're twentysomethings dressed in holiday finery, having a few drinks before a party.

Mongelli's been bartending for 12 years, and she's had plenty of people pour their hearts out to her. But not at Lefty's. They're more likely to do that at, well, weddings. She is a managing partner in a mobile bar, so she's seen a few morose reception guests. "It's not even when everybody says their 'I do's,'" she says wryly. "It's when the music starts and they are sitting with the kids' table, seeing their friends in couples and everybody's slow dancing and everybody's crying because it's so beautiful. And the lights are dim and you just start reflecting and thinking about how you've gotten yourself in that position." When people do spill their guts, they tend to do that in the comfort of their neighborhood roost—a place like the Night Light. "But it's definitely not a bar like this," she says.

Down the bar, Alex perks up. "I came here when I had an issue!" she says.

"Actually, that's true," Mongelli says in amazement. Alex says she'd just had a fight with the person she was dating, and when she came into the bar Mongelli instantly knew what was up. "I sat down and I didn't say anything and she's like, 'What's wrong?' And I was like, '*Nothing!*'" Alex delivers the recalled "*Nothing!*" in a high-pitched quaver.

"Well, you were also crying. It wasn't hard to figure out," Mongelli says equably, bringing her a Jameson.

OK, so maybe nowhere is immune to the pain of romantic loss. And in the course of one evening, the people sitting at this bar—a bar where, I repeat, happy people come to be happy—deliver what is essentially a doctoral defense on social pain.

Here's Mongelli making up a Belfast Car Bomb for Steve and Maureen, who immediately start discussing what Eisenberger would call thinking about the mental states of others. "The worst is when you keep thinking over and over and over how you could have done better," Mongelli says.

"What happened?" agrees Maureen, a cheerful woman with long curly hair and a bright turquoise sweater. You stew, she says, on "what you could have done to make it change."

"I shoulda did this. I shoulda did more," puts in Steve, her backward-baseball-hat-wearing boyfriend sitting farther down the bar.

"And you can't change it!" Maureen says. "*You cannot!*"

"And you all of a sudden start to paint a good light on that person and bad light on yourself," says Mongelli, bringing over the drink.

"You turn it around," agrees Maureen.

"You're like, 'Oh, *I can't believe I did that*,'" says Mongelli.

If they were psychology grad students giving a presentation, by this point they would be showing us a slide of the dorsomedial prefrontal cortex.

Here are Alex and Russ on which form of pain lasts longer. "Physical pain, it hurts at first and it goes away. Mental pain, I think it sticks with you and that's what makes someone grow," Alex says.

"Physical pain is so momentary. That's all it is. It's the moment it happened and then it just passes," agrees Russ. "The emotional one sticks with you forever. Well, it can stick with you forever if you allow it to." (Cue slide of Eisenberger's memory reliving study.)

And why might social pain last longer, Alex? "It fucking cuts you," she says bitterly. "Your heart, I feel like it kind of breaks a little. Like a piece flakes off and you're like, 'Well, there goes that.'" (Click to the slide of the International Association for the Study of Pain's official definition, the part about an emotional experience described in terms of tissue damage.)

Over to Mongelli, Steve, and Maureen, who are trading symptoms of heartsickness. "I've had a broken heart so bad that I literally thought I was having heart attacks, when it was my anxiety causing the physical pain," says Mongelli.

"You're hurt, you're sick to your stomach, you don't want to eat," says Maureen.

"You don't want to be around your friends," Steve adds.

"It's almost like the flu. Do you agree?" Maureen asks. "Your back hurts. You don't feel like eating. You're just stressed out."

(Now around this point, some of you might be thinking these symptoms actually sound a lot like depression or social anxiety. Stay tuned for Chapter 8, where we'll go deeper into the loop between the perception of emotion and physical symptoms. But for now, just think of Mackey and Eisenberger and their distributed network of pain sensory areas. Or imagine these folks at the bar showing you a slide that says "pain matrix.")

Here are folks we'll call Brian and Tara, who just walked in after a day at a conference, and are now puzzling over whether drugs like alcohol work interchangeably for different types of pain. "I've had a drink because I was physically in pain and also emotionally in pain," says Brian. "Like, I would drink to not think about emotional pain. And then I would drink because it helps numb physical pain. So I don't know—I mean, they could have been doing the same thing and I was just interpreting it differently."

He is not kidding about using alcohol to numb physical pain. When he was in college he separated his shoulder, but didn't have medical insurance, he says. "So instead of going to the doctor, I just drank a couple days in a row until the pain had subsided," he says, as Tara nearly falls off her bar stool in horror. And similarly, he says if he was "in a social situation, where if you just got broken up with or this crush that you had clearly wasn't into you anymore," he'd head to a bar where he could talk with his friends. "Getting drunk helps," he adds. "It helps you kind of forget." (Swap alcohol here for acetaminophen, and you basically get Nathan DeWall's hypothesis.)

My point here is not that these ideas are so obvious that anyone sitting in a bar can figure them out. There's a world of difference between the gut feelings of people being pestered by a reporter and a growing collection of peer-reviewed studies. The point is that these ideas are resonant enough that they've created patterns of behavior that we've all observed, that we can easily recognize in ourselves and in each other. In fact, these ideas are so resonant that we have set aside special places where we go to talk about them. Some of them are extremely private, like a therapist's office. And some of them are located in the middle of a crowd, in the back of the restaurant, somewhere between the roast beef and Marilyn Monroe. We recognize these patterns not because we are all psychologists and smarties, but because we are all painfully human.

YOU MIGHT BE WONDERING why we evolved a way to feel horrible.

Mackey puts it this way: "Pain is so wonderful because it *is* so horrible." Pain is, essentially, a protector. "Pain is probably one of the most primitive, most teleologically conserved experiences that we have" as animals, he says, "dating back to single-cell organisms kajillions of years ago. It serves to keep us away from danger or things that may be dangerous for us, that represent harm or threat." Think about the consequences of not being able to feel pain, he points out, like people who are born with what's called "congenital insensitivity to pain." "While we might think that's wonderful—and it makes for good TV, by the way, and good movie characters—the reality is incredibly tragic," Mackey says. "These kids don't feel when they put their hand on a hot stove and they are burning their hand. They don't feel if they step on something sharp and cut themselves, and so oftentimes they end up dying of overwhelming infections. Often they chew their tongue out because they can't feel it. Often they have to wear mitts, because if they scratch their eye with their finger they won't feel that it's painful." Pain even protects us from dangers we don't consciously register, the tiny twinges that prevent us from sitting too long in one position or from overloading our joints when we're moving. "And these kids don't do that, so they will end up with terrible, terrible arthritis at an early age because they don't get the feedback on shifting positions," he says.

"So pain is so wonderful *because it is so bad*," Mackey continues. "It only becomes a problem when it becomes chronic. That is where pain changes from being something that is highly protective, highly adaptive, and usually self-limited into something that is pathological that serves, to the best of our knowledge, no teleological survival basis. And it is relentless and in essence becomes a disease in and of its own right."

Eisenberger and DeWall argue that the same protective mechanism is at play when it comes to social pain. "As a human species, and a mammalian species, we are very dependent on other people," Eisenberger says. "As infants we are completely dependent on somebody else to provide us with food, to provide us with protection, with warmth. Later on, as a very social species, we rely very much on the social group. And so it may be very adaptive to have some signal in place that alerts us to when we are separated from our group of others or when we are losing those close social bonds."

Her group's theory, she says, is that over evolutionary history, "this social attachment system may have literally piggybacked right onto the physical pain system, borrowing that pain signal to alert us not only to when our bodies are in danger of being damaged, but when our social relationships are in danger of being damaged." In other words, when you're doing something that threatens your social well-being, it should hurt. "It makes sense that evolution would have selected for this deeply ingrained response of aversiveness to being rejected," DeWall says, "because when you get rejected, your chances of surviving and reproducing plummet."

"Obviously, it's not fun to feel social pain. It's not fun to feel rejected or hurt," Eisenberger continues. "But we think that there would be some negative consequences to never feeling bothered or hurt by losing a close relationship or being excluded. So we think that this is actually an adaptive signal for maintaining close social bonds." To be without it would be dangerous, because you would be oblivious when your actions were negatively affecting the group, causing others to withdraw from you. In this scenario, the egocentricity and lack of empathy of sociopathy might be the equivalent of being born without an ability to feel physical pain.

As with physical pain, social pain needs to be horrible enough to prompt you to immediately change your behavior. "What the pain signal essentially does is it grabs our attention. It makes us unable to focus on other things that we might want to be focusing on, like getting something to eat or taking a nap," Eisenberger says. "It forces us to deal with whatever problem is at hand."

But it's not yet clear whether the pain response observed by these researchers is specific to personal rejection, or if it is actually a broader response to *any* perceived threat. Dr. Ian Lyons, an adjunct professor at the University of Western Ontario, put a new spin on this research when he tried a similar experiment in 2012—with math. He wanted to see if a different kind of threat—having to do math problems—would activate a similar neural area. As he points out, most of the research in social rejection has portrayed the pain reaction as an evolutionary preference favoring the ability to protect social bonds. But complex mental arithmetic is, evolutionarily speaking, a recently acquired cultural skill with no immediate survival benefit. "What we're showing is, 'Look, here's a response that looks really similar to the social-rejection-equals-pain response, but we're doing it in a domain that can't really have evolutionary origins,'" says Lyons.

In Lyons' experiment, subjects with differing levels of math anxiety were asked to solve problems while inside an fMRI scanner. And indeed, the high-anxiety participants showed more activity in their brains' pain areas, specifically the bilateral dorsal posterior insula, an area associated with processing bodily threats, than when doing an equally difficult verbal task. Intriguingly, Lyons' subjects showed a pain response *before* they had to do the math problems, not while doing them. That means "*thinking* about having to do math is painful, which really speaks to this anticipatory, psychological interpretation element of it, rather than the actual math," he says. "Which, if you think about it for a second, kind of makes sense, because math can't hurt you. As scary as those numbers may be, they can't actually jump off the page."

His results open up some interesting questions. One is whether this pain response is learned through experience, rather than evolutionarily driven. Another is that it's hard to tell exactly why the math-phobes are anxious about math. If their pain stems from dreading a hated task, it could be part of a generalized threat response. But if the pain stems from anxiety about performing poorly, perhaps because they are worried peers or teachers will think badly of them, that could be an indicator of social distress.

Some researchers have even argued that the pain matrix could be more correctly called the "salience matrix," a network of areas that react to anything that captures your attention. "I don't know that I buy that point of view," says Eisenberger. But she does think it is possible that the social rejection response is part of a more generalized one for many threats. "So the idea would be that this is sort of a broader neural alarm system of some kind that could respond to lots of different survival-related threats," she says. "Physical pain is one of them. Social pain is one of them."

THERE IS ANOTHER THING YOU CAN GET AT A BAR—support from friends. And if you flip the story of John Nackley and Kerry around, and focus not on what they are talking about, but on what they are doing, that is what you get. It is also what you get if you think about Lisa Mongelli and her crowd at Lefty O'Doul's, all of them talking about pain, but in actuality having a night of friendly bonding. And it turns out that the bookend to all the research probing the awfulness of social pain is flipping the process around to see what happens when you try to alleviate physical pain with love or friendship.

Mackey's lab ventured onto this turf in 2010 with a study led by Jarred Younger, then a postdoc in his lab, now teaching at the University of Alabama. They recruited college couples in the throes of "passionate love," or what the study defined as the first nine months of a relationship. (I know that's very short. Sorry.) Passionate love is the period of time, Mackey says, "when you are incredibly attracted to the person you're in love with; you feel an intense emotional draw towards them; you have highly focused attention; you think about them all the time; when you are near them you feel great; when you are away from them you feel terrible. And doesn't that just sound like an addiction? Because it is."

The brain systems involved with both passionate love and addiction are linked with reward and craving. These include parts of the dopamine pathway, like the nucleus accumbens and ventral tegmental area. Dopamine is what Mackey calls "our feel-good brain chemical." Dopamine, he says, "makes me feel good when I have my dark chocolate in the afternoon, it makes people feel good when they have their Starbucks latte, it makes people feel good when they take a hit of cocaine, and it makes our young Stanford undergrads feel great when they are in love with someone." And—just as important—these areas are involved with analgesia, or reducing pain, once you get your reward.

For this study, the researchers asked one member of every couple to get inside the scanner and undergo a series of painful stimuli, ranging from high to no pain, while looking at a picture of their partner, looking at a picture of an equally attractive acquaintance, or doing an attention-demanding word task, a way to test if mere distraction is the true analgesic (something like "Think of a vegetable that isn't green" or "Think of a sport that isn't played with a ball"). Each time, the patients rated how much pain they felt. And as far as pain perception goes, Mackey says, "Love works great! It's a wonderful analgesic." When exposed to moderate pain, people reported 44 percent less pain when looking at their partner than when looking at an acquaintance. For extreme pain, the reduction was lower—about 12 percent, on par with the reduction produced by the distracter task.

Mackey is careful to point out that this painkilling effect isn't specific just to the rush of new love. It could be caused by anything that activates the brain's reward center and dopamine system. "I mean, if you took a hit of cocaine, probably the same thing would happen," Mackey says. "But the good news is that passionate love is a lot more socially acceptable."

In 2011, Eisenberger's lab did a similar study, but with more established couples, who had been together for an average of two years. The subjects were given mild or high heat pain while lying in the fMRI and viewing a picture of their partner, a stranger, or a neutral object. Once again, the people who were shown images of loved ones reported feeling less pain than the other subjects. Eisenberger was startled that such a subtle reminder of the loved one was producing an effect. "That almost seems kind of magical," she says. "Why would being shown a picture of your partner reduce your feelings of physical pain?"

This time, she didn't think the answer could be chalked up to dopamine and the rush of falling in love. "Behaviorally, these [studies] look like they are producing the same thing, our long-term couples, his more short-term passionate love couples," she says of the Stanford group. "But the neural underpinnings might be a little different." With longer-term couples, she thought, the answer lies in the ventral medial prefrontal cortex, an area that is often studied in connection with fear research, because it's associated with detecting safety and turning down the body's threat response. The presence of your partner would cue this part of your brain that you are safe, and diminish your pain response. And indeed, in her study, people showed more ventral medial prefrontal cortex activity when viewing their partner's picture—especially those in longer-term relationships, or who rated their partner as a significant source of their social support. When activity in this area went up, activity in the other pain regions went down, and people reported feeling less pain.

In another twist, they've also found that caring for somebody else makes *you* feel less threatened. In an experiment with twenty couples, they asked women to lie in the scanner while their male partners, directly outside, received a painful electric shock. When the women held their partners' hands—a supportive gesture—as the men received the shock, the women showed increased activity in their own ventral striatum and the septal area. Both brain areas are involved in reward, and the latter in fear reduction. The women who showed more activity in the septal area showed reduced activity in their amygdalae—which control fear response—while they were soothing their partners. (This didn't happen in the control conditions, in which the women held squeeze balls instead of a person's hand, or only touched their partners when they were not in pain.)

Eisenberger thinks this makes evolutionary sense, too, because caregiving relates to the survival of offspring. "If you have offspring that you need to take care of and you are facing some threatening situation, you don't want to be so freaked out that you are running away from your offspring. You need to run *towards* them," she says. "We may need some kind of threat reduction mechanism in place when we are in that caregiver role. And so that may bleed over to these situations when you are just sort of helping other people."

Most of us don't spend a lot of time watching our loved ones get electric shocks, but we do see them suffer in other small ways, going through break-ups or professional rejections, being left out by the other kids. "I think about this sometimes with schools," Eisenberger says. "Obviously you are not allowed to physically hurt somebody else, but there are no real rules for emotionally hurting somebody, for saying they can't play." Eisenberger thinks that if we had a better appreciation for how much social pain is like physical pain, we would be more tolerant and sympathetic. We would respect that people can be hurting, even in the absence of obvious signals like blood and broken bone. And it might bring some relief to the sufferer, too, to know that they're not just imagining their pain. "Sometimes people can feel like, 'Maybe it is all in my head? Maybe I should get over it, like my friends are telling me,'" Eisenberger says. "Just knowing that it's real can be helpful, too."

And there is one last thing to say about pain: It ends.

Here's one more story from Lefty O'Doul's. Steve and Maureen met in junior high school. "He was my first love," says Maureen. But eventually they went their separate ways. Steve married someone else, was married a long time. A few years ago he went through a painful divorce. It was a rough time, he says. He withdrew from friends. He didn't want to go to work. He couldn't sleep. "Oh man, I didn't even live no more, honestly," he says. He steals a glimpse over at Maureen. "Until she came into my life."

After decades apart, their paths crossed again. They started dating long-distance. Then they moved in together. It's been seven months now. It's not always simple, starting over as adults. "It takes time. It's hard," says Steve. "But if the love is there, I think it's going to work out."

EIGHT

....

Emotion

IT'S NIGHTTIME IN THE CULTURE and Emotions Lab at Georgetown University, and it's time to make a volunteer sad.

Tonight's subject, Participant 57, is a woman in her early twenties, with long blonde hair, pale blue eyes, and a smattering of acne. She's sitting in a bare room on a red chair, a big sheet of white paper taped to the wall behind her. In front of her is a video screen so she can watch a short film as a camera records everything she does.

As cultural psychologist Dr. Yulia Chentsova Dutton adjusts the camera's gaze from the observation room next door, student Alexandra Gold walks into the room with the subject and begins to gently attach electrodes to her. A heart rate monitor belt goes across her ribs. Two respiration sensors, which record how frequently and deeply she inhales, go on her chest. Two skin conductance sensors, which measure sweat—an indication of emotional reactivity—go on her fingers. Then Gold briefs the woman on her task: "Your job is to watch the film, and after you are finished, I will come back into the room to give you some follow-up measures asking you about how the film makes you feel. It will ask whether you felt any emotions in response to the film and how intensely you felt those."

That sounds pretty straightforward, but it actually contains a clue about what the researchers are after: The subject has been very subtly instructed to pay attention to her emotions. Not all of the subjects will be cued this way. Some will be told to pay attention to their bodies. Others will not be cued at all. The researchers want to know which people will more strongly perceive the internal emotional aspects of sadness—feelings like gloominess or anxiety—or the physical aspects of sadness, bodily changes like the welling of tears or a lump in the throat.

And there's another difference. Half of the subjects in this study are European Americans, and the other half Chinese Americans, most of them born outside of the United States. The lab's real question is about something bigger than a onetime verbal prompt. It's about whether all the prompting you've received in your life—your whole culture—directs how you experience emotion.

Their theory is that just like language or your past experiences with scents and foods, culture works as a soft biohacking apparatus: In a world of infinite information, it teaches you which patterns to see, what is salient to you, where to direct your attention. Depression and sadness have dozens of symptoms, both emotional and physical (or what researchers term "somatic"). But we don't pay attention to all of them, or at least not equally. The signals that prompt you to realize you are sad, or the red flag that warns you it's time to reach out to a friend or professional for support, can vary depending on what your culture considers the most important or bothersome.

Chentsova Dutton is the daughter of an astronomer, and she likes to think of emotional perception as similar to the way we observe stars in the night sky. "There is this huge variety of stars, and it is completely impossible for us computationally to actually notice all of them. But our culture provides us with an idea of which constellations are important," she says. "I know Orion and I know which stars go into that. I am able to draw it, I am able to detect it in the night sky, because I have a model. And so, similarly, our cultures provide us with constellations of emotions, of physical sensations, that are important for us to pay attention to."

Her study builds on a long tradition of work—including at Stanford University, her alma mater, and at Concordia University in Montréal, Canada, in the lab of her collaborator Dr. Andrew Ryder—showing that culture influences emotional perception. Chentsova Dutton and Ryder study negative states—depression, sadness, anxiety—while at Stanford Dr. Jeanne Tsai, who pioneered much of the lab-based work in culture and emotion, studies happiness. In more than a decade of lab and clinical studies, they have shown that when processing sadness or depression, people who grew up in Chinese cultures pay more attention to bodily sensations, while Euro-Americans pay more attention to emotional thoughts. People learn to pick out symptoms, they think, like we learn to pick out stars.

It's important to make it clear that these labs are not arguing for a sort of racial essentialism, that Asian people innately do Thing X while white people naturally do Thing Y. In fact, they are making the opposite claim: behavior is not inborn. Instead, our culture teaches us cognitions and behaviors that affect how we perceive our own internal states, and how we express them to others. And cultures are not monolithic, Chentsova Dutton points out. "In a given culture, there may be some shared ideas that everybody in this culture is going to be familiar with and has some exposure to—but how I react to those ideas might be quite different than how you react to these ideas," she says. "So we are not the same, but we have this shared system of coordinates." Indeed, some researchers in this field deliberately use the plural term "cultures," even when discussing a single nationality (like "Chinese cultures" or "Chinese cultural contexts"), to signify that cultures are man-made and internally diverse.

Culture doesn't have to mean national background, although nations are useful to study because they are so large that you can find patterns within them, despite variation among individuals. Much of the first wave of research focused on North American and Asian, particularly Chinese, cultures, because it grew out of a preexisting body of clinical and cultural psychology work there. But this research is now moving into other parts of the world: Chentsova Dutton's lab is also working in Ghana with researcher Vivian Dzokoto, and in Korea and her native Russia, while others are doing similar work in Mexico, Israel, Turkey, and several western European nations. Tsai's lab has tried slicing study groups along religious lines, and their labs collaborated on a project exploring gender differences.

So far, much of the work on negative states has been done with therapy patients, but for tonight's study Chentsova Dutton is working with non-depressed women, who will be only temporarily saddened by a short movie. She wants to see what happens when you tell people from different backgrounds to pay attention to one facet of sadness or the other. Will telling a European American like Participant 57 to pay more attention to her emotions (presumably the culturally normative social script for her) heighten her emotional reaction? Would telling her to instead pay attention to her body cause her to be less sensitive to emotion? Will the reactions from the "no cue" control group break along cultural lines?

Chentsova Dutton will record reactions in three ways: by using the physiological data from the sensors the woman is wearing, through her filling out a variety of self-reports, and finally by analyzing the micro expressions on her face. She uses a method called FACS coding to analyze very quick involuntary muscle movements that convey emotional information, even if the person is trying to repress or conceal how she feels. (Its full name is the Facial Action Coding System, and it was developed by UC San Francisco psychologist Paul Ekman.) This is an incredibly detailed analytic method—every minute of video takes a half hour to decode, she says. But it will allow her to observe small changes the participant may not be aware of, little movements of the lips, quivers of the chin, crinkles at the corners of the eyes. The lab uses three measures because people are surprisingly bad at judging their own physiological states—their self-reports can be wildly out of sync with the other metrics. So after rolling a short neutral film that looks like a screen saver to create a baseline for Participant 57's reactions, it's time for the sad movie. And it is really, *really* sad.

It's an ink-and-watercolor short, chosen because it uses no language and the animated line drawings could be anyone. As Parisian-style accordion and piano music plays, a father and daughter ride old-timey bicycles past a clump of cypress trees down to the edge of a lake, where they embrace. The father walks toward a waiting rowboat, turns back, and runs to give the girl one last enormous hug before rowing away. The girl pedals off down a long empty road, leaving her dad's bike behind. The girl returns another day, watches in vain for her father, then rides away. Wind and flying leaves indicate seasons changing, as the girl keeps coming back, older each time. The bike is always there, but never her dad. She returns as a grown woman with her husband and kids. She returns as an old woman, the road covered in snow, birds flying by in an empty sky. We can see that the lake is drying, its shoreline receding. She returns one last time as a woman so old she can no longer pedal her bike or prop it to stand up. By now the lakebed has become a marsh. She walks out into what was once water, cutting through tall grass. In a clearing, she finds her dad's boat, empty. She climbs inside, lies down and curls up.

If you are bawling right now, join the club. About a third of the subjects have cried so far. Participant 57 has stayed dry-eyed, and mostly just looks thoughtful. But Chentsova Dutton, watching on camera, says her face has shown a few subtle signs of emotion: She's furrowed her brows, had a slight lip quiver, and at one point was pressing her lips together, which people sometimes do to regulate emotions. "Her skin conductance went down. Her heart rate went down, which is consistent with the fact that somebody is sad, because sadness is deactivating," she says, checking the physiological metrics on the screen. But does the young woman perceive herself as sad? Gold heads into the room with a sheaf of forms for Participant 57 to fill out. We'll find out in a minute.

AS WITH THE STUDY OF PAIN, the study of emotion forces you to ask a deceptively simple question: What is emotion? And it turns out that there are a few different ways that researchers think about this.

Jeanne Tsai, the Stanford psychologist who was Chentsova Dutton's doctoral advisor, points out that the field has a classic definition: Emotion is a feeling state that involves physiological, subjective, and behavioral responses to a meaningful event. In other words, she says, "There's something that's happened in the environment and you need to respond to it." From an evolutionary perspective, it's a mechanism that lets you rapidly react to a reward or threat without thinking too much about it.

Evolutionary biology tends to emphasize very quick reactions, and indeed, emotions typically last only a few seconds. But, Tsai points out, people also have moods, which last longer, and personality traits, which can last a lifetime. And not all emotional reactions are to external stimuli—you can react to a memory, or an imagined scenario. So an evolutionary definition doesn't cover *everything*.

As taste experts do, some emotion researchers categorize emotions by sorting them into basics, which are considered innate and universal. (In this case, each is linked with a distinct facial expression, not a chemical receptor.) There's no one agreed-upon set, but most include happiness, sadness, disgust, fear, anger, surprise, and contempt. As with basic tastes, each is associated with an adaptive behavior. "So fear allows you to flee from a predator; anger allows you to attack something," Tsai says. The basics are then overlaid by complex emotions, like shame or pride, which some researchers consider more learned and influenced by culture and language.

Tsai leans toward a more subtle, four-quadrant way of sorting emotions in terms of arousal (or excitement) and valence (positivity or negativity). "So you can have high-arousal positive states like excitement, enthusiasm or low-arousal positive states like calm and peacefulness," she says. Or you can have the opposite: "high-arousal negative states like fear and nervousness, and even anger, and more low-arousal negative states like dullness, sluggishness." This is called a dimensional model, because it sorts emotion states along two or more axes. There are many variations, but arousal and valence are very commonly used.

In 1949, Yale University neuroscientist Paul MacLean introduced the idea of the "visceral brain," which he later redubbed the "limbic system," a set of structures key to emotion that mediate between external sensory stimuli and one's internal state. These included much of the area formerly called the rhinencephalon (or the "nose brain," a testament to the intertwining of emotion and olfaction) plus the amygdala, orbitofrontal cortex, and hippocampus, which he thought correlated all sensations. Modern researchers dispute much of his schema, arguing that some of these areas are now known to have other functions, and other emotion-related structures are left out. Nevertheless, the limbic system name stuck, and MacLean got many notable structures right, particularly the amygdala, which receives early input from the sensory systems and triggers emotional and threat detection reactions.

So emotion is triggered from within and without, and we feel it within and without. And just as pain researchers do, emotion experts wrestle with a language issue: whether the body-related words we often use to indicate emotions—a burning anger, being sick with jealousy, feeling light with bliss or heavy with depression—are metaphorical or actual sensations. Chentsova Dutton thinks they are both, although some cultures teach you to separate them more than others do. "Initially, they are indeed a very much interconnected stream of information that our brain and mind are processing and making sense of," she says. "The separation of this sensory emotional stream into separate information about how *I* feel and how *my body* feels has more to do with constructs that our culture is providing for us about how to categorize what is happening to us."

In fact, she says, "In many languages it is really impossible to talk about emotions without talking about sensation." In her international research, she often gives people questionnaires asking them to describe their emotional states. "In Russia and Ghana," she says, "oftentimes they come back asking questions that indicate to me that to them the domain isn't carved that way, that thoughts and emotions and sensory perceptions aren't distinct to them." But English offers both bodily metaphors for emotion *and* a separate emotional vocabulary—words like "joy," "tranquility," "melancholy," "anomie." American culture also generally encourages people to articulate and share feelings. Because this seems so natural to Americans, she says, when she asks U.S. students to fill out a page-long inventory on how they are feeling, most do it in under a minute. By contrast, she says, "Asians, Russians, and particularly Ghanaians find it excruciatingly difficult to do this task. Ghanaians take about 20 minutes on the same task, and Russians take anywhere between 5 and 6, so much, much slower."

So your language and home culture teach you what you should pay attention to. Then the mere act of paying attention to something reinforces it. Let's say something delightful happens and you feel a little tingle on your skin and think, "I am happy!" Now you are going to read skin tingles as a sign of happiness, because you have made a connection between the word and a bodily sensation. And this works on the group level, too. If everyone successfully uses certain emotion-related words or bodily symptoms to communicate feelings— like signaling stress by talking about a headache, or disaffection with the word "meh"—now that connection is socially reinforced. And finally, because our brains and bodies work together, the more you think about physiological symptoms, the stronger they get. "If I pay attention to the fact that my heart is beating, I am fairly likely to make it beat even more," says Chentsova Dutton. "My attention creates this feedback loop."

Andrew Ryder, her frequent collaborator and a clinical psychologist at Concordia, offers the panic attack as a perfect example of a physical and emotional feedback loop in action. Say you notice you're anxious. "Maybe you are worrying about something," he says. "Maybe you had four cups of coffee rather than two—something that might raise your heart rate a bit, make your palms a little sweatier, might make your chest a little tighter." If you are prone to concern about these subtle changes, that might make you *more* anxious. Now your heart beats even faster, your palms sweat a bit more, your chest tightens further. "As you're noticing that the potentially problematic thing is happening, you are simultaneously fueling that thing," says Ryder. "You go around the circle a few times: Now your heart is racing and your chest is really tight and your palms are super sweaty. And now you're freaking out because you're having a panic attack."

You can clearly see the role of culture in this loop, he says, because not just anything can fuel a panic attack—and what does varies around the world. You would never have a panic attack based on, say, a pre-sneeze tickle in your nose or a weird pain in your pinkie, even though these might be just as surprising and noticeable as sweaty palms, because no culture associates sneezing or pinkie pain with impending danger. Yet people in North America become alarmed by a racing heart or chest pain because they're feared as precursors to heart attacks. "If you didn't have the cultural belief that chest tightness and so on is related to a heart attack, then the loop wouldn't really be able to form," Ryder says.

Now imagine your neck hurts, not your chest. For North Americans, that might sound unpleasant, but not alarming. Yet in Cambodia, researchers have observed that neck pain can trigger panic attacks because of a belief that the rising of blood and *khyâl* (or "wind," an air believed to flow through the body) to the neck will cause blood vessels to rupture. Chest tightness and sore necks are both symptoms of stress; during a rough moment, you might have both, along with a half-dozen other indicators of anxiety. But your culture has taught you which one to beware of, and so you'll attend to it over the others, and possibly magnify it to the point of panic. "So, if I drink too much coffee, my neck might get stiffer because my muscles get tense, and my heart might start racing," says Ryder. But as a North American, "I'm just not going to care that my neck gets tense. It might bug me. I might rub it. But there is no potential there for it to form into a kind of self-reinforcing loop that will cause me to freak out. So I will focus in on the heart and the chest issue."

You can find examples all over the world where people are sensitive to distress indicators that outsiders might never notice. Chentsova Dutton likes to note that in Russia, people pay attention to quick shifts between hot and cold because of a long-standing belief that it makes people sick. As a result, she says, few offices use air conditioning, drinks are rarely served with ice, and people are careful to monitor their surroundings for drafts. Or, she adds, when she was in Ghana, where malaria is endemic, she was surprised to find college students casually asking each other whether they'd been tired or feverish lately—just something people living in a West African climate find salient to monitor about their bodies.

The same feedback loop also works with social anxiety. Say you're at a party, and you're worried about how you appear to others, so you're constantly scanning for signs of rejection, Ryder suggests. Maybe your anxiety makes your hands tremble a bit, so you get a glass of wine to have something to hold. Now say the person you're talking to briefly glances away. Have you bored them? Did someone more interesting walk into the room? "These are the kinds of things that when I think them, make my hand shake," says Ryder. "That's OK, I've got my glass of wine. I'll hold the glass of wine tighter. But it turns out that holding the glass of wine tighter actually helps your hand to shake *more*." By now you're afraid the shaking will become visible, or that you'll embarrass yourself by spilling the wine. And ironically, as you monitor yourself even more closely, the ruder and less responsive you may appear to your conversation partner, causing them to withdraw and giving you more cause for worry. Feedback loop achieved.

Or take depression, which is linked with the twin demons of fatigue and sleeplessness. The more you focus on how tired you are, the more you worry about not being able to get a good night's sleep. So even if you feel exhausted, Ryder says, "you go to bed with a feeling of dread, like, 'I'm going to have yet another night where I get 90 minutes of sleep.' The very act of anticipating of this, of course, is making it worse."

There's another feedback mechanism overlaying emotional perception, which is that feelings don't happen in isolation, but within social contexts. For example, depression is often stigmatized. A person might fear judgment from their peers, or be treated poorly by them, for talking about being depressed or showing symptoms like constant fatigue. And that stigma worsens mental health problems in a way it can't for physical problems. "If you break your leg, and then everyone thinks you are a loser for breaking your leg, it doesn't make the leg *more broken*," Ryder says. "But if people are all shunning you because you are depressed, that in itself is a depressing experience." Similarly, a constant need for reassurance or help with mood regulation might strain your social networks, making you feel even more alone. "So you have this nasty cycle of the worse I feel, the more I need. But the more I ask for help, the more rejected I feel," says Chentsova Dutton.

In this extremely complex system, where feelings, physiology, and interpersonal relationships overlap and feed back into each other, what you learn to pay attention to is important. It governs how you understand yourself, and how others understand—and treat—you. "The universe is very, very complicated, and we only can possibly devote cognitive resources to a fraction of it," Ryder says. "Culture becomes really important because it provides us with these guides."

But the kicker is that our cultural guides tend to be invisible to ourselves. They are so deeply familiar that we assume whatever we perceive is unlearned and universal, that what sadness is like to me is what it is like to you. So if you really want to know how the world feels sad, then you've got to recruit some very patient volunteers, train a camera on their faces, hook them up to a bunch of monitors, and then, ever so slightly, break their hearts.

CHENTSOVA DUTTON IS BACK in the observation room, cuing up video of a woman who came in a few weeks back. This is Participant 37, she says, rolling a close-up shot of a young Chinese American woman with freckles and blunt cut bangs. For a while, the woman just sits silently, staring ahead as she views the animated short. People's emotions tend to break through at certain moments, Chentsova Dutton says, as the film moves through the cycles of the girl's return to the lake. The emotions get more intense with each return, and if people are going to cry, they do it when she finally climbs into the boat.

Participant 37 seems to be trying very hard not to cry. She keeps touching her hand to her face, although it's unclear whether she's scratching her nose or reacting to welling tears. "It seems like she is trying to regulate by biting on the lip," Chentsova Dutton says, "sort of putting a brake on her expression." The woman gives a few other pre-cry signals: She's blinking; her lip corners are starting to turn down; she sniffles a bit. Her lip begins to quiver, and she keeps biting it to keep back tears. But now it's a losing battle. "There the chin is starting to shake a little bit," Chentsova Dutton says, and immediately two big tears stream down the woman's face. "There we go. There we go," Chentsova Dutton says in a soothing voice, nodding her head at the screen.

This subject was cued to think about her body, and indeed, according to her self-report, she's almost exclusively paid attention to it. On a 0 to 8 scale rating how strongly she felt a long list of sensations, she's given the highest marks to physical ones like tearfulness, changes in breathing, speeding heart rate, and feeling warm, along with the more cognitive category "memories triggered." But she has given herself very low scores for emotional categories like gloominess and distress, and a zero for sadness itself. Yet even if she wasn't actively feeling sorrow, she seemed aware that she was *thinking* of something sad. When asked to describe how she felt after watching the film, she penned, "Sad thoughts of my friends before. I miss home."

So far, says Chentsova Dutton, this is a typical result: Chinese Americans are indicating much more awareness of how their bodies are responding, and European Americans are more likely to report that they subjectively feel sad. But there are exceptions, and when Gold returns from doing the post-film survey with Participant 57, it looks like she may be one. As a European American who'd been cued to pay attention to emotion, their hypothesis predicts that she'd be quite sensitive to feelings. But her results are mixed. Although she'd displayed some subtly sad facial expressions, her self-report is markedly analytical, largely about how much she enjoyed the music and the drawings in the film, and her efforts to suss its meaning. "I don't really feel a strong emotion. It was a little sad," she wrote, scoring herself a zero in that category.

"Wow!" exclaims Chentsova Dutton, reading over her results, which show that while she gave herself the highest scores for feelings, they were for positive ones like attentiveness, clear focus, and contentment. "So she almost has like an aesthetic pleasure in it," she muses. "It *is* a beautiful film." Maybe the drop in skin conductance they'd noticed earlier wasn't because the woman was becoming sad, but because she was becoming *calm*. In emotion research, that's just how it goes. People are not, individually, predictable. "There is always a whole range, and some people react or not," Chentsova Dutton says. "So the point is to detect significant differences relative to the fact that there is this noise in how people are responding."

And as a broader culture, patterns show through. By the time the lab finishes their analysis the next spring, it will become clear that Participant 57 was an outlier. Overall, the lab found that verbally cuing people to pay attention to body or emotion had no effect; instead, the differences broke down along cultural lines. While just about everyone reported feeling sad, the Chinese Americans more often reported that they were tearful or experiencing other bodily sensations. The European Americans showed larger increases in skin conductance—an unconscious bodily response linked with emotional reactivity—while the Chinese Americans reported more consciously sensed physical symptoms like muscle tension, rapid heartbeat, and a lump in the throat.

Overall, the researchers concluded, the breathing and heart rate data did not show that the Chinese American participants had a heightened physical reaction to the film—if anything, the European American participants had a slightly stronger one. "This finding supports the idea that culture shapes emotional responding through the interpretation of experience, rather than through psychophysiological response," they concluded. So you read emotion top-down, rather than bottom-up. Your cultural preconceptions influence what you perceive more than your bodily state does.

CHENTSOVA DUTTON'S STUDIES ARE BUILT on the Tsai Lab's concept of "ideal affect"—not what you feel, but what you *want* to feel. Tsai calls ideal affect "a desired state, an unconscious or conscious goal or state that people work to attain," and she realized that this might be different from people's actual affect, or the way they truly feel in the moment. "Culture obviously influences both," Tsai says. "But we think that culture influences how you *want* to feel more than how you *actually* feel, because culture teaches us what's good and desirable and virtuous."

So while everyone aspires to happiness, she believes, people vary in what kind of happiness they chase. Her lab has focused on studying American and Chinese or Taiwanese participants, and she's found that, in general, East Asians report that they value calm, serene happiness states, while Americans tend to value high-excitement ones. (Remember the four-quadrant sorting method: both have a positive valence, but different arousal levels.) They've found that these cultural differences can be seen as early as preschool age, and they hold even after controlling for an individual's temperament—their overall level of emotional reactivity.

To find how these values are conveyed and reinforced in ordinary life, her lab has studied the smile—the way we indicate inner happiness to the outer world. They have analyzed everything from children's storybooks to fashion and news magazines, Facebook profiles, and photos of corporate CEOs, all showing that Americans are consistently more likely to portray high-wattage smiles, while East Asians are more likely to show calmer ones. (Her lab also uses FACS coding for measuring the emotional content of smiles, using cues like whether the mouth is open, if the teeth are showing, and whether there are crinkles at the corners of the eyes, which some consider an indicator that the smile is genuinely felt, rather than forced.)

These ideal states—rather than actual states—influence how we organize our emotional lives and much of our behavior, Tsai says. Say you're feeling blue and want to cheer yourself up; if your ideal state is an excited one, you might go running to achieve that high. If your ideal state is calm, maybe you stay home and read a book. Maybe, she suggested in a 2007 paper, it even affects your culture's drug of choice; the stimulants cocaine and amphetamines are more abused in the United States than are depressant opiates like heroin, which are more commonly abused in China. What you value also has to do with how your culture urges you to interact with others, she argues. If your culture encourages personal independence and having influence over others, you tend to value energetic, excited states. If your culture encourages group harmony and adapting to others, you value calm states. And as we'll see a bit later, her lab is also exploring whether your ideal state is an unconscious motivator for subtle judgments you make about the people around you.

Chentsova Dutton's work branches off by studying negative states, with the idea that these, too, are desirable in their own way. Negative emotions are important, she says, because they are warning signals—so important that we can generate them internally. "Unlike animals, we are able to produce emotional states through thinking," she says. "A zebra is only going to be horrendously stressed if a lion is chasing it, but my students can get themselves in the same state by worrying about their midterm." And, she points out, it's much easier for us to think ourselves into worry than into happiness. People rapidly habituate to the positive but are easily alarmed by the negative. "Humans have evolved to be exquisitely sensitive to threats," she says. "If you are highly reactive to threats, you are going to be probably in better shape for passing your genes on."

American psychology tends to portray negative feelings as dysfunctional, she says, but "in many cultural contexts, people would think of them as quite important and ennobling and useful." When Chentsova Dutton interviews Russians, they often say that they want sadness to be part of their lives—and their children's lives. "They say that it helps their relationships, that it helps them be more empathetic towards others, that it helps them problem solve and be more objective. It helps them remember that they are just human, not be too hubristic, sort of as a counterweight to excitement and happiness," she says.

One of her recent projects explores whether achieving your ideal affect can help you perform an attention-demanding task, even if your ideal state is negative. She tells her participants they must solve a puzzle, but first asks them to choose an emotion they would like to feel as they work, which she'll help them achieve with a film or music. "Americans very consistently say they want to be happy and Russian very consistently say they want to feel sad in that situation," she says—and now she's testing whether feeling that emotion does indeed help people focus.

In two other collaborative studies with Tsai, she concluded that people defy their culture's ideal affect when they are depressed. "European Americans stop showing their emotions," Chentsova Dutton says. "Asians do something that is equally contrary to the Asian norm. They instead become kind of labile; they become overly emotional." Perhaps a depressed person just doesn't have the energy to mount a culturally appropriate response, she says. Or perhaps the person becomes *more* depressed because they are culturally out of sync and drawing the disapproval of others.

She and Ryder are now exploring how these differences in perception manifest themselves in the clinic. Since 2002, Ryder has been studying depressed and anxious patients in both Canada and China. "What we've been doing is trying to study how these different cultural frames direct our attention to particular experiences that turn them into *symptoms*," he says, or signals so bothersome that they prompt someone to seek help.

Ryder points out that there have been some awfully unflattering, and sometimes outright racist, attempts at explanation since Westerners began studying Chinese psychiatry patients around the 1970s. Some of the early theories presumed that the European emphasis on the mind is the norm, and drew on a psychoanalytic tradition that considered a focus on the body, or somatization, an immature defense mechanism. Therefore, patients who did it must be emotionally unsophisticated or repressing feelings. (The term "Chinese somatization" was coined to indicate a deviation from the norm; Ryder suggests the term "Western psychologization" would be equally plausible.) Others, perhaps more sympathetic, suggested people were avoiding the stigma of mental illness by focusing on bodily symptoms, particularly in a Communist regime where admitting to feelings like hopelessness about the future could come off as an implied critique of the government.

Ryder and his colleagues believe that what you perceive comes down to its relevance to you, what is most meaningful to communicate your distress and what consequences you expect—everything from assistance to being shunned. In a 2008 study working with patients at clinics in Toronto, Canada, and Changsha, China, he found that in open interviews, in which the patient spontaneously spoke about their problems, and in more structured ones, in which a clinician asked about standard depression symptoms, Chinese patients were more likely to mention bodily problems, and Canadians negative thoughts. The Chinese patients were much more likely to complain of, for example, fatigue, sleep problems, and weight loss, while the Canadians of hopelessness, loss of interest and pleasure, and low self-esteem. That wasn't true for every symptom they examined, though—both groups were about equally likely to mention pain and dizziness. (Ryder's group is now working on a follow-up study that will compare data collected in Korea and China, looking for other demographic differences, including between rural and urban populations.)

A complementary study, also looking at which distress symptoms people describe, completed in 2014 by doctoral student Eunsoo Choi in Chentsova Dutton's lab, compared Koreans and European Americans who were asked to write to a therapist or a friend about a situation that made them angry. Then they rated how effective they thought they'd been in communicating their state and provoking empathy. Choi analyzed the texts, finding that Koreans used more bodily references than Americans when recalling the incidents, and that using somatic words made disclosing the experience more satisfying to them, and made them expect more sympathy from the reader. (Americans didn't have any such expectation.)

In a second part of the study, to test whether readers really would be more sympathetic to bodily references, Choi asked Korean subjects to read some of the narratives, one describing a frustrating job search and one an upsetting work situation. But she altered the conclusion of each. Some ended with mentions of physical distress like headaches, trouble eating, and hair loss. Others ended with reports of feeling depressed and "down." And indeed, Korean readers responded more sympathetically to the physical descriptions, more often indicating that they pitied or wished to assist the person. It's an example, Choi concludes, of how your culture guides the strategy you choose to get support or consolation from others.

Thanks to his own clinical experience, and what he hears from his collaborating physicians, Ryder believes it's rare for any patient to complain *only* of emotional or physical symptoms. It's more a matter of emphasis, and often a different take on cause and effect. Most of the Chinese depression patients he's studied also mention some emotional symptoms—it's just that they see the root problem as a physical one. "So it's like, 'Are you depressed, anxious, guilty, feeling bad about yourself?'" he asks. "And they're like, 'Yeah, you know, I haven't slept properly in about a year and it's screwing my work up and my wife is losing her patience now. So of course I feel those things. But those things aren't the disorder nearly as much as the sleep problem is.'" On the other hand, somebody raised in North America might view the feelings as the true problem and the rest as side effects. "When you are worried all the time, you don't eat very well, you don't want to get out of bed because you feel hopeless, and so on," Ryder says.

And that's important, Ryder says, because the whole point of this work is to help clinicians work better with patients. If you understand why they might be bothered by one thing more than another, you can avoid making a bad diagnosis, or implying that there is a "right" way to feel depressed, or ignoring concerns because you don't find them relevant. "Not being understood by your therapist is also depressing," Ryder adds. "So if I am too quick to cram them into my own cultural box, then it's not just that I won't understand them as quickly—I could be actively making them feel less understood and more alienated." What matters to the therapist should be what feels real to the client, and how they expect those feelings to be interpreted by others. "I want to know what they are perceiving," Ryder says, "and then how they perceive other people perceiving them."

STRESS AND HELP SEEKING don't just play out in a doctor's office; they're part of everyday life. Ryder's lab believes that here, too, you can see cultural patterns at work. So once again I'm watching by camera from an observation room, although in this case, we're at his lab in Montréal, and the room we're viewing is much more comfortable than the deliberately bare Georgetown lab. This time, it's a sort of ersatz student lounge with pale yellow walls and puffy maroon chairs facing a coffee table.

Psychology student Biru Zhou is about to put two volunteers through a demonstration of her doctoral research experiment. Overall, she says, "I'm looking at how social anxiety differs in different cultures." She's particularly interested in a syndrome known in Japan as *taijin kyofusho*, or an extreme fear of offending or hurting people. It's an interesting form of social anxiety, she says, because rather than being focused on the self, it is focused on others. In this case, adds Ryder, the person doesn't fear being personally rejected as much as they fear disrupting group harmony—and that's significant in Japan and China where the well-being of the group is often considered the priority. In this case, Zhou's going to test differences in how stressed people ask for and give help.

This part of her study looks at how pairs of friends—both of either Chinese or Euro-Canadian descent—behave when given a daunting task. She's recruited fellow Concordia psychology student Yue Zhao, as well as Momoka Watanabe, a cultural psychiatry student at nearby McGill University, to stand in as today's subjects. (Normally, both would be of the same ethnic heritage, but because this is just a demonstration, the friends are Chinese and Japanese.) As they settle into their maroon chairs, Zhou reveals the job she's about to give them: Draw downtown Boston on an Etch A Sketch. Or rather, one of them has to draw it while the other one helps.

This is, frankly, impossible.

"*Exactly,*" says Ryder.

"But that's not what we're looking for, whether they can do it or not," adds Zhou. The drawing is just a stressor. The real test is to see if there will be a cultural difference in how people interact as they deal with what Zhou calls "the unsolvable."

As at the Georgetown lab, participants here are wired with heart rate monitors, fill out self-reports afterward, and are filmed, although this time the observers are watching for tiny behavioral and verbal cues that indicate support seeking or giving. They're looking for direct requests for help, like asking aloud, and indirect ones, like complaining or acting frustrated. They'll also look for indications of support, like giving the drawer advice, boosting their self-esteem, or making them more physically comfortable, as well as for "negative behaviors" like discouraging them or being distracting. So this time, there are two sides to the emotional signaling: what the first person does to show distress, and how the second responds.

Zhou readies the Etch A Sketch and the picture of Boston her subjects must copy, and heads into the yellow room. "They will be so shocked," she says a bit gleefully.

The lab keeps the study random by assigning the drawing task to which-ever subject happens to sit in the "secret chair," and today it's Watanabe. Zhou tells her that she will have ten minutes to copy the picture she's about to see. Then Zhou turns to her second subject and tells her she can help, but she can't take over. "You have to remember this is her task, OK?" she instructs.

Zhou flourishes the picture of Boston, as both of her subjects burst into laughter, and exits, telling them, "On the count of 3—1, 2, 3, go!" Watanabe begins drawing intently as her friend watches. From the other room, the re-searchers point out that the two women are sitting closer together, that they giggle frequently, perhaps as a tension release. As the helper, Zhao lends a hand literally—holding the wiggling Etch A Sketch against the coffee table—and by giving advice and moral support. "You are so good. It's amazing!" she says as their giggling intensifies.

"OK, so now which direction do I go?" Watanabe responds.

This is exactly what the researchers are looking for—dozens of interactions showing people reading each other's emotional cues. After ten minutes, Zhou interrupts the women and assigns them a new task. Now they must sit in front of a webcam, and Watanabe must pretend that she is introducing herself to a person she would like to know better. Again, her friend can assist, but can't do the task for her. If you think drawing Boston on an Etch A Sketch is hard, try improvising a ten-minute on-camera speech to a stranger. "There's a rea-son why this one is always second," deadpans Ryder. It's tougher because it's personal, self-revelatory.

Watanabe valiantly launches into an effort to describe herself, and deflates after about four sentences. Her friend leans over and whispers a suggestion. The two of them keep going; every time Watanabe runs out of ideas, she glances sideways like, "What now?" and together they manage to find something new to talk about—Montréal's food scene, cooking, how to get around town.

As she watches their tape, Zhou says she expects that both groups will ask for and receive help equally, but the Chinese subjects will more frequently ask indirectly, because asking overtly could disrupt the group, or place a so-cial burden on others. "We are very much attuned to harmonious relation-ships, so the support seeking and support provision should be there at all times," she says. "It's almost like, 'I need help but I shouldn't say it, *because the other person should know it.*'" This is the positive side of knowing how to read your culture's social-emotional world; it tells you how to help. At a more negative extreme, it makes possible a social anxiety like *taijin kyofusho*.

When she finished her study a few months later, Zhou found she was par-tially right—but there was a twist. The Chinese participants actually directly asked for help *more* often in both tasks. But she found that among Canadian pairs, the more often the drawer indirectly asked for help in the first task, the more likely the friend was to behave negatively—by critiquing or discourag-ing them—in the second one. Among the Chinese pairs, indirectly asking for help in the first task later made the friend *less* likely to behave negatively. Zhou concluded that the study design might have led to the unexpected results. After all, the subjects were explicitly told to help, which likely made it seem obligatory, and asking for help justified, rather than disruptive or self-serving. But indirect support seeking was better tolerated by Chinese partici-pants, she concluded, because it aligned with a cultural expectation that requests should be made without seeming to burden others.

Additionally, she points out, previous studies only looked at overt support seeking between strangers, not friends—and she wanted to know if people behave differently in close relationships. In this case, she points out, it looks like they do, so "the unexpected results are only 'unexpected' in relation to previous studies."

BY NOW, YOU KNOW THAT when psychologists have isolated a behavioral pattern, and want to understand its underlying neural mechanisms, they turn to the fMRI scanner. That's why on a bright spring afternoon, Tsai Lab doctoral student BoKyung Park and lab manager Elizabeth Blevins are helping a young woman get comfortable on the narrow bed they are about to slide into the magnet. In partnership with Tsai's husband, Stanford psychology and neuroscience researcher Brian Knutson, the lab is now exploring how we "write in" information about other people's emotions, rather than "reading out" our own. In other words, rather than influencing how you perceive your own internal state, how does your ideal affect help you interpret the feelings of others?

Once Park and Blevins have tucked their volunteer in with blankets and foam pillows, they gently slide the coil over her eyes and return to the obser-vation room. Park cues up what the subject is about to see: images of faces, so close up that you can't see past their chins or hairlines. They're a mix of male and female, East Asian or European American faces, all smiling with different degrees of excitement or calm. After seeing each face, the subject will respond to one of two questions about personal traits: How good of a leader is this person? Or: How familiar is this person? (Here, "familiar" means how much that person is like someone they might encounter in everyday life.) Each time, they'll respond on a scale of 1 to 4.

All of the subjects in this test are European American or East Asian women, and all of them will fill out surveys afterward to gauge their ideal affect. The researchers will correlate their neural activity with their trait ratings and the results of their questionnaires. For the white participants, Blevins asks, "If they see a very excited smile, will they view them as a better leader than somebody with a calm smile?" And vice versa—will the Asian subjects see the calm smilers as better leaders?

Now how widely someone smiles might seem like a pretty trivial metric for leadership. But over the past several years, Tsai's lab has repeatedly correlated ideal affect with how people judge the competency and trustworthiness of others. And this has consequences in the real world, where people from different cultural backgrounds mix all the time, and often have to make "gut" social judgments about whom to hire or promote, whom to trust or befriend, or whom to elect. "The question is, on what are people making those judgments?" Tsai asks. She thinks those gut feelings arise from the expressions you see on the other person's face, specifically from how well the emotions they display mirror the state you value. "If you come from a culture that values excitement, then you are going to see an excited face as more friendly, more trustworthy, and ultimately a better leader than if you come from a culture that values calm," Tsai says.

If you consistently have a mismatch between a minority group's values and what the cultural majority expects, she thinks, that's how you get phenomena like the "bamboo ceiling," or the difficulty Asians working in Western cultures have rising past middle management. Tsai believes they're not perceived as being as animated and enthusiastic as their corporate counterparts, and therefore not as dynamic leaders. Even President Barack Obama—who spent part of his childhood in Indonesia—might be a victim of that mismatch, she muses. He's often been criticized for his constant calm, considered by some to be too passive for an American politician.

So as a first test of the leadership idea, her lab compared politicians' smiles. Using over 3,000 official headshots of legislators from ten nations, they found that the more the nation valued excitement, the higher the percentage of politicians bearing broad smiles. (To measure a nation's ideal affect, they used national surveys of college students.) The United States, along with Germany and France, came out on top for politicians showing excited smiles, with China, Hong Kong, and Taiwan on the bottom. Mexico and Japan were in the middle. Interestingly, South Korea ranked among the European nations for politicians displaying broad smiles. Possibly, says Tsai, it's because South Korea has a large Christian population. In previous studies her lab found that Christian texts and worshippers emphasize excitement states more than Buddhist ones do—and that's the dominant religion in the other East Asian nations.

And here you can see how the snake eats its tail. On a gut level, you feel most positive toward the candidate whose emotional demeanor seems the most appropriate or pleasant to you. Smart politicians know how their culture expects them to act. Once elected, legislators have handlers who help them choose a photo that makes them seem statesmanlike—whatever that means to the surrounding culture. And that reinforces voters' ideas about how a leader should look. "It's what cultural psychologists call the 'culture cycle' or 'mutual constitution,'" says Tsai. "Culture is created by people, but then people are influenced by the cultures that other people create."

These emotional judgments can show up in all kinds of life choices. Tamara Sims, a former student in Tsai's lab, did a study in which people were asked to choose a doctor based on profiles in which the physician endorsed either an active lifestyle or a calm one. The participants generally chose the doctor who reflected their own emotional values. In a follow-up study, participants randomly assigned to a "virtual doctor" whose behavior was either excited or calm were more likely to adhere to health recommendations if the virtual doctor reflected that person's own ideal affect. So if you value excitement and you get an exciting doctor, Tsai says, "You are more likely over the course of a week to do these health behaviors like drink more water, go to sleep earlier, not eat two hours before you go to bed, to walk more." Similarly, the lab asked medical students to judge patients, and found that the doctors-in-training said they believe people who reflect their own emotional values will be better patients and that they expect to be more effective with them.

We probably don't even realize that emotion has a role in making what we suppose are logical decisions: We give our preferences other names, like deciding who has charisma, or is likeable or seems like a "good fit" for the organization, Tsai says. "Perceiving other people, it's so automatic you don't even think culture influences it. You just think you know this person is friendly," Tsai says. "Or you think that person is a jerk." These assessments, she continues, "are so automatic they seem that they're *real*." But they are, of course, in your head. And Tsai would like to know exactly where. So in the fMRI observation room, Park taps on the controls, and up pops the first of the faces. Over the next 15 minutes, the woman will judge dozens of them. As she works, Park records her brain activity.

The lab is particularly interested in the activity of three areas, each one representing a different theory about which neural mechanism underlies the subject's judgments. "Maybe the Caucasians just think those with excited expressions are better leaders because they are matching with their self-concept," Park says. The lab calls this possibility the "cognitive mechanism," or the idea that people identify with a target who reflects their ideals. If that's the case, Park says, they expect to see increased activity in the medial pre-frontal cortex, which is associated with identity and self-related information processing.

A second explanation is a visual one. Perhaps people who really value ex-cited faces pay more attention to them and process them differently than calm ones. In that case, when the person is viewing someone who matches their ideal, the researchers expect to see more activity in the fusiform gyrus, the area of the brain that responds to faces.

But the third possibility—and the one their data eventually bears out—is that it's an affective or emotional mechanism: People find faces that match their ideal more pleasant. That's signaled by greater activation in the ventral striatum, which includes the nucleus accumbens. As you may remember from the experiments with love and pain relief, this is part of the brain's dopamine system. Receiving the type of smile you crave is a tiny reward.

And so if this is right, it means that what your culture teaches you to pay attention to guides not only your perception of your own emotions, but how you interpret the emotional world around you. You learn to read other peo-ple, just like you learn to read your own internal state, just like you learn to read stars.

From the neural to the social level, we must cope with the computational vastness of the information stream, the unending universe of what there is to know. The brain handles this overflow by guiding our attention, using pa-rameters shaped by soft biohacking forces like language, experience, social relationships, and cultural reinforcement. We don't perceive everything; we can't. Yet the world feels whole, synchronous, and orderly, without missing parts or limits, even though we only peer at the universe through the tiny keyholes of our senses.

But humans are engineering creatures. Throughout our history we have used technology to extend our powers and enhance our experiences. And this process is only speeding up. As we saw with the development of neural implants and brain-machine interfaces, we are increasingly able to tinker with our own sensation and perception. While these first-generation tech-nologies are primarily restorative, developed to return some degree of func-tionality to those with medical needs, we are on the cusp of an era in which we may be able to give ourselves perceptual abilities and experiences beyond what nature allotted us. And we create faster than evolution. It took nature millions of years to evolve the eye. It took man only a few decades to build the retinal implant.

The next generation of perceptual technologies won't simply assist; they'll be designed to augment and alter our versions of reality. In doing so, they'll bring devices closer to the body and more routinely into daily life. In the final act of this book, we'll see what happens as the sci-fi future starts to merge with the present, and the line between person and machine begins to wear away.

PART THREE
· ·
Hacking Perception

NINE

· ·

Virtual Reality

BRUCE JOHN, A SOFT-SPOKEN GUY in baggy jeans and a scruffy gray sweatshirt, is taping electrodes onto the soldier who sits patiently on the raised platform in front of him. Skin conductance sensors go on the soldier's left middle and pointer fingers. Two for heart rate go just inside the elbows. The soldier is already wearing a respiration belt under his sternum, timing his breaths.

In a month, he will deploy to Afghanistan.

In five minutes, he will go there virtually.

The soldier (no names; they're confidential) is tall and broad shouldered, wearing khaki fatigues and boots laced tight around the ankle. He's a member of the National Guard, part of a unit called the Special Operations Detachment–Korea, based out of Buckley Air Force Base in Colorado. The members are mostly part-time soldiers who also work civilian jobs, led by Colonel Kenneth Chavez, who in his civilian life has spent more than 35 years on the Denver police force. He's already deployed to Afghanistan, Iraq, and postearthquake Haiti. But for a third of the group, this will be their first tour of duty. Only a handful of them have jobs at home—police or hospital work—that regularly expose them to the kinds of stresses they are about to face.

These soldiers will mostly be working with their Afghan counterparts to prepare them to run operations after the United States reduces troop levels abroad. They aren't expected to face combat. But, Chavez says, they will still risk danger. They will travel in convoys, which means the threat of improvised explosives. They will work on bases, which can be targets for attack. And Chavez knows that even if his soldiers are physically safe, they could come home carrying emotional freight. "One of the things that can cause people a lot of stress after a mobilization or combat tour is guilt," Chavez says. "You could have survivor's guilt. You could have guilt of inaction—you should have done something. It might have prevented someone from getting hurt, or saved them. It could be preventing injuries to civilians. You might have witnessed something that weighs heavily on your mind. And you carry that for the rest of your life."

Chavez wants his troops to be resilient, and that's why volunteers are being wired up in the next room. They're taking part in a study led by Dr. Albert "Skip" Rizzo, a researcher from the University of Southern California's Institute for Creative Technologies. Rizzo's a psychologist, and his healing tool is virtual reality. For more than a decade, he has been working with the military to treat post-traumatic stress disorder, or PTSD, a condition in which the brain's natural "fight or flight" stress reaction continues to engage long after the danger is over. PTSD is a real problem for the military: A 2014 Congressional Research Service report estimated nearly 119,000 cases had been diagnosed among Iraq and Afghanistan veterans since 2000.

Rizzo's Virtual Iraq and Virtual Afghanistan simulations use a goggle-like head-mounted display to create an all-around sensory experience that plunges soldiers back into their wartime experience, allowing them to relive their most anxious moments under the guidance of a therapist. Repeating those moments is meant to rob them of their power to terrify. This technique is based on the much older concept of exposure therapy, which asks the patient to either literally reexperience a stressor or imagine it aloud to their therapist. "People can confront what they fear or what makes them anxious," says Rizzo. "They can do it in a safe environment and nothing bad happens. All of a sudden, the fear starts to extinguish."

Virtual reality, or VR, offers an incredibly powerful way to wrap yourself in that imaginary landscape. It is perhaps the earliest and most potent example of a computer technology specifically designed to alter sensory perception. As the field has matured, it's provided fascinating insights into what happens as we merge our senses with the electronic world. It's shown that we react physiologically, emotionally, and intellectually to the virtual world as we do our own, even carrying those feelings and behaviors into ordinary life. Around the world, VR labs are running that ball in opposite directions. Some work to create ever-more-fantastical bodies and terrains, defying physics and biology to see how readily the brain adapts to novel existences. Others, like Rizzo's, focus on the hyperaccurate. He aims to create a sensory environment so real it tricks the mind into healing.

The watchword in VR is "immersion." For a simulation to work, you must feel completely inside of it, reacting to that world naturally. When the technology was nascent in the early 1990s, that feeling was hard to come by. VR was almost purely audiovisual, rendered through head-mounted displays the size of diving helmets. Images were cartoonish, and lags in tracking and rendering—locating your moving body's position and redrawing the world around you—made for a nauseating ride. Today, it's sleeker, faster, and way more multisensory. Rizzo's group employs a bass shaker beneath the platform on which the soldiers are seated. It vibrates so that when a bomb goes off, the floor trembles; when they turn on their Humvee's engine, the chair rattles. When Rizzo works in his home lab, participants patrol the virtual streets holding a prop weapon that mimics the look and feel of a rifle. A machine pumps odors into the room: rotting garbage, diesel fuel, sweat. Rizzo wonders whether they should add a heat lamp to mimic the desert sun. (He's stumped on taste, though. "I can't imagine how we could use it," he says. "We could put a spoonful of sand in people's mouths or something?")

Today will be the first test of Rizzo's newest idea: Can you make soldiers resistant to PTSD *before* they head into combat? This time, the soldiers won't relive trauma. They will *pre*live it. Maybe if they virtually handle situations before facing them in real life, they'll fare better. Maybe they'll still be angry, sad, or anxious—but those feelings won't become debilitating. As Rizzo likes to put it, "We're trying to put ourselves out of a job on the back end by doing a better job on the front end."

Chavez signed on to test this new idea, dubbed the STRIVE program (Stress Resilience in Virtual Environments), because his time on the police force convinced him that surviving stress can be productive, if it's handled well. "I've seen numerous officers that have gone through critical incidents and have either suffered from that or have grown," he says. "So I believe in post-traumatic stress disorder. And I also believe in post-traumatic growth."

The darkened room where John is running the simulation is littered with the detritus of a military unit about to deploy: a table with piles of compact discs offering lessons in Dari and Pashto. Department of State guidebooks for working in Afghan provinces. A pyramid of bottles of a bright red nutritional drink, which claims to be fruit punch but has the distinct flavor of sweat socks, which everyone on base is doggedly drinking nevertheless.

"I'm going to check the software really quick just to make sure we're getting the data," says John. He watches the soldier's respiration rate on the display. The soldiers have already had their blood drawn to be analyzed for stress indicators like cortisol, dopamine, and norepinephrine. A second set of readings will be taken when they return. Maybe, says Rizzo, they'll find connections between how anxious people subjectively feel and what this data reveals about their biological states.

"All set," John says. "I'm going to put this head-mounted display on you." It's a slender set of white Sony goggles, not much bigger than the kind used for skiing, held on by Velcro straps. Then he adds the "occlusional shroud," a band of cinched black fabric that goes over the goggles, sort of a tube top for the eyes. "This is really just to block out the light and peripheral vision," John says. Next he hands the solider a video game control pad. The goggles will track his position as he moves, but he'll need the controller to make some maneuvers and answer questions the simulation throws at him.

The solider will spend nearly two hours in this virtual world, going through six simulations that follow an Army unit in Afghanistan, what Rizzo calls a "Band of Brothers" storyline. One day it might be 30 parts, to be experienced over weeks like a TV series. For now, it's shorter, but it's not going to be pleasant. Bad things are going to happen, to the soldiers and to the people around them. Because it is a teaching tool, every so often the simulation will halt and a virtual mentor will appear, talking the soldier through the physiology of stress and offering advice on controlling anxiety. The solider will be asked to rate his stress level, or indicate his emotions. But for the most part, he'll be along for the ride, a member of the squad, seeing, hearing, and feeling as they do.

"I'm going to begin the scenarios here," John says. "There may be times where you become discomforted. This discomfort is not going to be long lasting. It will be like something you encounter watching a movie or playing a video game. If you become so discomforted you can't continue, please say 'Stop.'"

He presses a button. The soldier, sitting motionless and shrouded on the platform, is here. But now he is also *there*.

AS THE SOLIDER GETS GOING on his virtual journey, Rizzo is in the next room, typing clangorously into an enormous laptop, a red plastic monster that contains worlds; if you plug in headphones it can become a VR-lab-to-go. Rizzo gives the impression of being large in every way; he speaks in a jovial, rapid-fire New York accent and has hugs and handshakes for everyone. He wears scratchy work shirts and his long graying curly hair is tied back in a ponytail. He fondly refers to the colonel as "brother." The two have gotten along like a house on fire since the day Chavez saw a CNN piece on Rizzo's work and fired off an email asking the scientist to bring it to the base.

Rizzo doesn't hail from a particularly high-tech background. Twenty years ago, he was working as a clinical neuropsychologist, doing cognitive rehabilitation for people with brain injuries. He was frustrated with the tools of the trade: mostly dull workbook exercises, ineffective for patients who by definition have trouble paying attention. Then one day he spotted a patient—a young man who'd survived a car accident—out under a tree playing what was then the latest in entertainment technology: the Game Boy version of *Tetris*. Because of his injury, Rizzo says, it was hard to keep this patient on task for more than 10 or 15 minutes. "But here he was *glued* to this game," Rizzo recalls. And then it hit him: What if you could make therapy just as engaging?

That Christmas someone had given him *Sim City* for Nintendo, the ubernerdly urban planning game. Rizzo loved it, and thought his patients would, too. *Sim City*, he says, "was the ultimate executive function and training exercise. You have to plot a strategy, initiate it, monitor it, repair it, adjust it—all the things that we consider to be the components of executive function." And like the kid with *Tetris*, his patients glommed onto becoming virtual urban planners. "It wasn't like rehab for them," he recalls. "It was like playing a game."

The next bit clicked for Rizzo one day while driving to the gym. A radio station was playing an interview with computer scientist Jaron Lanier, whose company sold the first VR gloves and goggles. Lanier was speaking from a Japanese showroom where shoppers could use a virtual interface to design a kitchen. Rizzo couldn't turn the interview off. Sitting there in the parking lot, he realized that with virtual reality, you could not only create everyday environments, but control everything that happens inside them. That's perfect for cognitive rehabilitation, he thought. So before heading into the gym, he ducked into a bookstore and demanded everything they had on VR. "And they had two books," he deadpans, which he read between sets. The next day at work, his boss handed him a flyer for the first conference on using VR for people with disabilities. "And it was like, 'Oh my God, *people are thinking about this already!*'"

For the next few years Rizzo moved from one job to the next, angling for anything that would give him access to better equipment. In the early '90s, VR was a growth industry; it had its own conferences, magazines, and movies. (Witness: *The Lawnmower Man, Strange Days, Johnny Mnemonic*.) "All the cyberpunks were engaged in it," recalls Rizzo. "There were books coming out on homebrew VR," or building your own system with TVs and other odds and ends. But the bubble popped, partly because the promise of immediately available *Snow Crash*–esque worlds, filled with exquisitely rendered exotic landscapes, life forms, and high-speed adventures, was slow to materialize. Equipment was expensive and cumbersome. Internet connections lagged. (Remember: dial-up.) Edges looked choppy; people moved robotically; everything was pixelated. "The first time I put on a head-mounted display and navigated around in a virtual city, it was really just very lame—untextured geometric forms, the navigation was awkward, the head-mounted display was heavy, and I ended up getting stuck in a building wall," Rizzo recalls.

But the academic research Rizzo was doing relied less on cinematic razzle-dazzle than on clinical utility, and he was eager to experiment with the medium. He built a spatial training program in which you had to mentally rotate 3-D blocks. He worked on a virtual classroom for kids with attention deficit disorder. Then in 2003, as veterans began returning home from the Gulf, Rizzo got interested in doing something that would help them, too.

Rizzo wasn't the first to think of re-creating wartime for soldiers with PTSD. His work builds on the results of a long collaboration between Dr. Barbara Rothbaum of Emory University and Dr. Larry Hodges, then a computer scientist at Georgia Tech, who pioneered exposure therapy in virtual reality. Originally a form of talk therapy, the technique has long been used to help patients confront garden-variety fears: heights, public speaking, spiders. Under the guidance of a therapist, you are gradually introduced to the thing you fear. First you talk about spiders, and then you see pictures of spiders. Eventually you work your way up to actual spiders. Virtual reality offered a way to make those fearful things seem real without them being real at all.

Before Rothbaum started working with VR in the mid-'90s, she was taking her patients out to confront their fears in real life, riding elevators with acrophobes and taking people who hate airplanes, well, onto airplanes. And that could be slow. "You have to leave the office. Usually it takes longer. You might violate patient confidentiality," she says. "In virtual reality, you can do it all in your office and in your 45-minute therapy session."

The first VR environments her group tried were for the fear of heights: elevators that participants would "ride," hotel balconies they could stand on, and what one subject dubbed the "Indiana Jones bridge," a rope-and-wood-slat span over a canyon. Those worked, says Rothbaum—not only did the pilot subjects report physical fear symptoms like sweating and shakiness, indicating they were responding to the virtual environment, but their anxiety decreased over time. Even better, seven of the ten subjects subsequently voluntarily exposed themselves to real-life height situations, even though the researchers hadn't asked them to. "That is what really matters," Rothbaum says. "It doesn't matter if you can get on a virtual elevator if you can't get on a real elevator."

For the next simulation, Rothbaum begged the computer science team for a fake plane. Fear of flying was a challenge to treat, because she'd have to go with her patients all the way to the airport and wrangle Delta into letting them onto a stationary plane. With the simulation, she says, "Like a real plane, most of the activity occurs out the window. You can see takeoff and Atlanta and the clouds. You can go through a storm and have the clouds get dark." That worked, too. Their next study found that virtual therapy was just as effective as exposure therapy on a stationary plane, and the majority of the patients were able to take a real plane flight within six months of the therapy.

Next, they thought of helping Vietnam War veterans. Researchers had noticed that people with PTSD often talk about "numbing out" right away, sometimes deliberately, blocking out anything associated with the trauma. This avoidance, a hallmark of the disorder, is likely making it worse. "About 70 percent of us will experience a traumatic event, but not 70 percent of us end up with PTSD," Rothbaum points out. Imagine getting back behind the wheel after you've survived a car crash, she says. Maybe at first you're jittery and hypervigilant, but if you keep driving, and nothing bad happens, you lose your fear. "But for folks with PTSD, they don't," says Rothbaum. "Partly it's because they avoid: They avoid thinking about it; they avoid going to the places that remind them."

Treating this disorder is different than treating a fear of heights or planes, where the dreaded disaster is imaginary. For trauma survivors, Rothbaum says, "their worst fears have come true." The person must resurface an actual memory. But by the mid-'90s, Vietnam veterans had been home for more than two decades. Those who were still struggling with PTSD and related problems—depression, substance abuse, relationship difficulties—were considered a treatment-resistant population, the hardest to help. Even though the therapy was novel, they didn't have much to lose.

In 1995, her group rolled out Virtual Vietnam, which simulated two common, emotionally potent environments for veterans: the inside of a Huey chopper, associated with arriving in a combat zone and evacuating the wounded, and a landing strip in a jungle clearing. (The landing zone made veterans so anxious—it reminded them of vulnerability to enemy fire—that the team had to add a hill to make them feel less exposed.) Once the veteran was inside, the therapist would ask him to describe his most traumatic memory from Vietnam, retelling it in the present tense. Meanwhile, the therapist controlled the environment to match the memory. For example, Rothbaum says, their first participant was a former helicopter pilot: "As he is saying he is landing the chopper, the therapist is landing the chopper. If he's saying, 'There is a lot of gunfire,' the therapist can produce gunfire."

At first, her team wasn't sure how immersed the veterans would become, or how they might react. Even though they were treating outpatients, they did their first clinical study in the inpatient wing of the Atlanta VA hospital in case anyone needed immediate psychiatric help. One part of the helicopter simulation involved male voices shouting "Move out! Move out!"—the research team was so worried that the soldiers would instinctively rip off the fantastically expensive VR helmet and throw it aside as they exited the virtual chopper that they tethered it to the ceiling. "And none of that was necessary," Rothbaum says. There were no meltdowns. No one threw the helmet. And overall, the therapy helped: Participants scored lower on standardized measurements of PTSD and depression after the study, and in follow-ups six months later.

And there were other extraordinary hints about how the mind could interact with computerized sensory input. Soldiers seeded the virtual landscape with details from their memories. "Somebody said that they saw tanks; we didn't have tanks in it," Rothbaum says. "Somebody said they saw the enemy. We didn't have the enemy in it. Somebody said they saw water buffalo. We didn't have the water buffalo in it." A bulldozer operator who'd had to dig a mass grave during the war mimed the entire memory using the mouse buttons they'd given him as pretend bulldozer controls.

In 2003, Rizzo launched his own demo for veterans of the Iraq and Afghanistan wars, hoping to reach not only a larger test pool, but younger soldiers who grew up with video games and might feel even more at home with virtual therapy. (He and Rothbaum are now frequent collaborators.) Rizzo had other arguments for why the idea was worth retrying. Virtual therapies—perhaps ones that could eventually be delivered by home computer or mobile apps—could overcome logistical barriers, like travel for vets who live far from cities, or long waits at overburdened veterans' services centers. And they might make seeking help easier for soldiers reluctant to ask for it in person. "Let's face it," says Rizzo, "in the civilian sector or anywhere, it's like, 'Do I want a sharp pointy stick in the eye, or do I want to go talk to a therapist?'"

But what turned Virtual Iraq from a demo into a real program—and for Rizzo into his main job—was that in 2004, the *New England Journal of Medicine* published an article showing that 11 percent of army soldiers returning from Afghanistan met the broad screening criteria for PTSD, and so did 18 to nearly 20 percent of soldiers serving in Iraq (depending on branch of service). "That got everybody interested in PTSD," Rizzo says. In 2005, the U.S. Office of Naval Research began funding Virtual Iraq, which was sent to 50 clinical sites for testing. In individual case studies and small clinical trials, Rizzo's group and others found decreases in PTSD symptoms, depression, and anxiety; in some cases, the subjects no longer even met the screening criteria for a PTSD diagnosis. While studies remain ongoing, these convinced Rizzo that virtual therapy is safe and effective for war veterans.

Rizzo thinks STRIVE can help even more people—not just predeployment soldiers, but perhaps civilians who routinely confront trauma, like emergency room personnel or natural disaster first responders. STRIVE hinges on the psychological principle of state-dependent learning: If you learn something in one emotional or physiological state, you'll recall it better in that state. "The classic example everybody uses in grad school is if you drink a shitload of coffee when you are studying for a test, you'd better drink a shitload of coffee when you are taking the test," says Rizzo. So if you can teach people about controlling stress while inducing stress, they might be better prepared to handle it in the real world. War is still going to suck, Rizzo says. It just might suck less.

THE SOLDIER IN THE COLORADO pilot study sits on a swivel chair in the darkened room, goggles over his eyes.

And now he's driving a Humvee across the desert, on a dusty road outside a small town. He sees his hands on the wheel, the vehicle's interior jogging up and down, the parched terrain rolling by; hears the clatter of the engine, the chatter of passengers in the back seat, heavy metal on the sound system. A narrator's voice has just told him that his squad is looking for a man suspected of supplying parts for explosive devices, and that the group's previous driver was wounded by IED shrapnel.

As the driver swivels his head, he can see who else is in the Humvee. In the passenger seat sits Corporal Soto, the squad leader, a no-nonsense type in a camouflage helmet and yellow-tinted sunglasses. Behind Soto is Backovic, a pale redhead, the group's designated smartmouth. Directly behind the driver, impossible to see because of the seatback, is the car's good conscience, a soldier named McHugh.

During the first long minutes of dull desert driving, the guys in the backseat bicker and tell dumb jokes and Soto keeps telling them to pipe down. Then the Humvee up ahead pulls off the road. Soto orders the driver to follow. The driver sees his hands lift binoculars to his eyes, and a dark spot on the road resolves into a clearer image. It's a body. It's a man's form, bloody, curled in the fetal position, wearing gray robes that indicate he's local, not military.

In the darkened room, the soldier visibly sits up straighter.

In the virtual Humvee, a fight breaks out over what to do. Acting as a sort of Greek chorus for the driver's own supposed thoughts, Backovic and McHugh argue for going to help, while Soto urges everyone to wait for the bomb squad.

And then, sickeningly, the man on the road lifts his head.

"Damn, he's alive," says a voice from the back seat. The tension in the Humvee escalates, as the soldiers argue over whether this is a trap. "This. Smells. *Wrong*," insists Soto, pointing out that the road seems strangely abandoned. But the guys in the back seat press their leader, asking if they can at least throw the man some water. "What do we do, just sit and watch him die?" asks one.

"Yes, we sit," says Soto. "And yes, we may watch him die. That guy may be innocent, or he may have a pound of C-4 stuffed up his ass. Maybe he's lying there just waiting for us to come help him, just waiting for us to get near. Then he or one of his buddies in that village hits a switch, and boom!"

While the animation isn't as realistic as film, it's vivid and finely detailed. The driver can see the grime marks on the windshield, the cracked plastic on the dashboard, flies buzzing around the body. As the Humvee's motor rumbles, the bass shaker lightly vibrates the platform. Since this is a mobile simulation, Rizzo's lab doesn't have all the sensory cues that he can normally use, but the dialogue attempts to convey them, as the soldiers complain about it being as hot as a sauna and smelling like one, too. At last, the bomb squad rolls up. A robot on treads heads toward the body.

Then the scene dissolves. Now there's a different man sitting where Soto was, leaning casually against the dashboard as he turns to face the driver. He introduces himself as Captain Branch, the driver's mentor, and speaks with forceful emphasis. He says he's going to help the soldier mentally and emotionally prepare, "so you come home safe and sound and not haunted by the stuff you're going to experience out here."

"Let's take today," the mentor continues. "You stood by and did nothing while a man bled to death on a roadway. How'd that feel? Wrong? Frustrating? Overwhelming? It sucked, didn't it? This kind of twisted crap happens all the time here. Your natural impulses are going to be challenged at every turn. What you've learned from every good and decent person in your life is sometimes going to have to go on the back burner. The right thing to do in San Diego or Charlotte or Where-the-Hell-Am-I, Idaho, could get you killed here. If you aren't prepared for how you're going to feel, if you don't know what stress can do to your mind when you walk into combat, you might as well leave your helmet and your battle gear in the barracks because you're asking for serious trouble. This is called resilience training. Think of it as your emotional gear."

Out on the road, the robot reaches the body. Suddenly an explosion rocks the bass shaker.

The soldier in the darkened room, who had begun to slump, sits up straight, silent but alert.

"And there's the truth," says Branch. "This time Soto was right. But maybe next time it won't be a trap and an innocent will die."

On-screen, the convoy begins to roll again.

THERE IS ANOTHER WING OF VIRTUAL reality research exploring how manipulating sensory perception can influence behavior, but through a very different kind of storytelling. I'm about to experience it while I'm in the shower.

Or, rather, my senses are in the shower, a white-tiled room where a steady stream of water pours over my head. My body is at the Virtual Human Interaction Lab at Stanford University, where it is wearing a VR helmet, my eyes peering into stereoscopic computer screens. At the back of the helmet is an accelerometer that tracks my head's rotation, on top is one of the five blinking infrared LED lights charting the position of my body, and from it I'm trailing a long set of cables, which lab manager Cody Karutz is making sure I don't trip over.

In 2008, when I first visited this lab for a magazine story, it was a nearly empty attic room in the communication department, and the rendering of the virtual worlds had a blocky, cartoonish quality. While impressive, you still got the feeling you were interacting with polygons. Today, the refurbished lab has the ultramodern feel of a high-end recording studio, all glass and fabric walls. Eight cameras track my movements to within one-tenth of a millimeter of accuracy. Speakers embedded in the walls and ceiling create a cloud of sound that follows me everywhere. Sixteen subwoofers below the floor shake it to emphasize motion. The visuals have more shading, detail, and subtlety; the water droplets falling over my head glisten in the light and move in realistic arcs.

Over the years, this lab has developed a body of work showing how virtual experiences, even very short ones, carry over into real life. Nick Yee, a former lab member, named this the "Proteus Effect" after the shape-shifting Greek sea god. Early experiments made nearly imperceptible tweaks to your avatar, or virtual body, and then tested how the changes shaped behavior afterward. Yee, now the author of a book on the Proteus Effect, found that people who were made slightly taller in the virtual world later became tougher negotiators in a real-life bargaining task. People who were made slightly better looking chose prettier partners when later asked to pick from a phony dating site. Those given unattractive avatars became more likely to lie about their height on the fake site. Former lab member Jesse Fox found that people exercised more often when an avatar built to look like them gained or lost weight according to their exercise level. She also found that women who spent time in avatars of themselves wearing racy clothing later scored higher on a scale of "rape myth acceptance"—the idea that victims cause or deserve the assault—than women whose avatars were scantily clad strangers. Perhaps, she concluded, it's because seeing yourself sexualized triggers guilt, or a defensive desire to believe that rape victims must have done or worn something you wouldn't have.

These subtle shifts in feeling are called "transfer," and now the lab is exploring how to harness them for social good. "I'm really trying to focus the resources, intellectually, of the lab to think about how we can use virtual reality to make the world a better place," says cognitive psychologist Dr. Jeremy Bailenson, the lab's founder, who takes a decidedly unstuffy approach to academic life. He surfs. He's into metal. He assigns William Gibson's *Neuromancer*, cyberpunk's classic VR adventure, to his beginning students every year and reads it along with them. He likes the idea of using the virtual to make things real, of playing with the boundaries between worlds. He wants to leverage what he calls "the catch-22 of virtual reality," or the idea that "from a perceptual standpoint, it can be close to indistinguishable from actual reality." In his written work, he often refers to the physical world as "grounded" reality, just one flavor of the realities to be experienced.

But instead of the tough verisimilitude of Rizzo's lab, Bailenson's VR is essentially a new form of magic realism. The demo Karutz is about to put me through has a lyrical, slightly surreal quality; by the time I rejoin him nine months later, the simulations will have become straight-up bizarre. "In the virtual world, there are no rules. You can break constraints and physics and reality," Bailenson says. "So what we're doing is designing simulations that can put people in wonderful or tragic outcomes that can really viscerally demonstrate the link between human behavior and outcomes."

That's why I'm in the shower, looking through a bathroom window at a very weird outdoor scene. There are two tables in the backyard. One holds a plate piled with coal. Behind the other table is me—or, rather, my avatar. If I'd participated in the original experiment, she'd be my doppelganger, but today I'm borrowing someone else's body. She's wearing a conservative blazer, wire-framed glasses, and a blonde bi-level hairstyle that falls to her shoulder on one side and is shaved up the other. Like most avatars that aren't being controlled by a player, she is semi-animated, cycling through subtle automated movements, shifting her weight and gaze back and forth. It makes her appear alive, if not exactly engaged, like someone perpetually waiting for a bus.

I jump as an overhead voice booms: "Your task is to look outside the window while you act as if you are showering. Notice the water coming out of the faucet and allow the water to get your body wet."

Out in the backyard, a lump of coal floats off the plate and over to my avatar. Who. Eats. *It*. She looks me dead in the eyes as she chews, a horrible brittle cracking noise. Then she coughs, the most baleful cough you can imagine, right into the crook of her arm.

"Now begin by washing your right arm," says the voice.

Another lump of coal floats off the plate as I wash. Chew, cough, cough. This time my avatar actually turns away from me as she hacks mournfully.

"Wash your right shoulder." Wash. Chew. Cough, cough.

The me inside the helmet is rapidly becoming unnerved. "I feel bad seeing this poor avatar eat this gross coal," I tell Karutz, or at least the point in space where I think he is. I am upset enough that I'm essentially ruining the premise of any VR experience, breaking the virtual fourth wall, undercutting the illusion by calling attention to the fact that it *is* an illusion.

The system isn't having it. Perhaps the tracking technology has sensed that I'm no longer looking in the right direction. "Please continue to look outside the window while you wash," the voice admonishes, and then adds, mercilessly, "Wash your right side."

"I can't do it any longer!" I beg. Karutz, who is kinder than the computer, pulls the plug. I lasted a minute in a simulation that is supposed to go on for six.

As Karutz helps me out of the helmet, he explains: The study was designed by lab member Jakki Bailey to study conservation for the Department of Energy. Each piece of coal represents the 100 watts of electricity required to heat and transport water for 15 seconds of showering. In the real experiment, participants washed their hands before and after at a sink that measured their water usage and its temperature. Some of the participants got the simulation I did, some saw the coal move between the two tables without an avatar eating it, and others simply saw a ticker of how much coal had been used. Those who saw the coal float by (with or without the avatar) used as much water as the others, but chose cooler water, suggesting they were trying to conserve energy.

The lab's other recent work explores building empathy for others or the natural world. In one study, the lab had participants fly, superhero-like, through a foggy downtown to rescue a diabetic child, to see if it would make them more helpful in real life. In another, the lab robbed participants of their ability to see green and red, testing whether it would make them more likely to assist a colorblind person. And there's the one Karutz is plunging me into now. I'm standing on a hillside, gazing up into the canopy of an enormous tree, listening to birdsong. I'm holding a chainsaw, or rather a controller that vibrates to match the saw's roar. I must cut down this tree, a task I do not want to do. But I saw, and the tree splits, and the crash seems to go on forever, and the birds stop singing, and I stand alone on the hill, and it's enormously sad.

In the real study, the test would have come after the experiment seemed to be over, and gauged paper waste. Some participants received the VR experience, but others only read a sensory-rich text and imagined themselves as the tree cutter. Afterward, researcher Sun Joo (Grace) Ahn, a former lab member now teaching at the University of Georgia, would "accidentally" knock over a glass of water and ask for help wiping up the spill. Then she'd count how many paper napkins people used. Both groups said that cutting the tree made them feel like their actions affect the environment—but those who'd had the VR experience used fewer napkins, a real behavioral change.

These real-life changes are significant, Bailenson believes, because they are a way to capture how "present" the person feels in the virtual world. "If the response is the same thing you would expect if one were in the physical world, then you are getting presence," he says. To him, it's no surprise that we think and feel in the virtual world just as we do in the physical one, and that we can't put those experiences aside when we take off the helmet. "The brain is not wired to differentiate virtual experiences from physical ones," he says. "If it looks like a cliff, I'm going to think it's a cliff. That's just what the brain does." The amount of time humans have lived in a world where one has to differentiate between electronic environments and real ones, he points out, is just "a speck in the evolutionary history."

He doesn't think there's anything unnatural about the ease with which we shed our physical lives for more ethereal ones. Humans have a long history of indulging in mental dissociation, from cinema, radio, and books to pretechnological consciousness shifters like storytelling and drug use. "If you look at the history of media, there's always been as much media as we can get. People enjoy pausing on their own life and transporting somewhere else," he says. "Even without media, people daydream constantly. So the mind likes to wander." VR isn't novel, in terms of releasing the mind from the body to go explore phantom lives; it's just *easier*. The brain has to do less work to construct an imaginary world. "You know," he says mildly, "You can dream on demand."

The bigger question, Bailenson says, "is when will you no longer know it's not real. And we're not there yet. But I certainly think the technology is moving very quickly." While his lab still uses traditional VR helmets, which run from $20,000 to $40,000, they also use much cheaper gaming equipment, like the Microsoft Kinect, a hands-free controller system for the Xbox, which tracks your body by bathing it in infrared light. Without you having to wear or touch anything, the Kinect copies your motions over to your digital avatar. The lab has also started creating scenarios for the Oculus Rift, a lightweight head-mounted display developed for home gamers. It can't replace an entire VR system, but the visual quality is good, and it may be the breakthrough product that brings virtual reality to the consumer market. (Facebook acquired its maker in early 2014, with both companies hailing VR as a computing and communications platform that would soon extend immersive experiences beyond gaming.)

Perhaps most importantly, Bailenson points out, the last decade of smart-phone use has done an excellent job of training us to walk around with one eye on the virtual, trailed by a cloud of online social interactions and tasks. Phone apps are not yet very immersive—you don't get a wraparound sensory experience—but they are very good at presence, or demanding one's atten-tion, making you feel like you are *there* rather than *here*. Bailenson thinks that as VR technology gets smaller and cheaper, it will become more embed-ded in everyday life, making it harder to separate our lives on the ground from our lives in the cloud. What happens, he asks, when you "can make Face-book feel like a frat party, or can make online gambling feel like Vegas, and you're still doing it *while in the presence of physical others*? How does that change the world?"

Before I leave the lab, Karutz puts me through one more demo to illus-trate the increasingly lush world of VR, how easy it's becoming to adapt to the dream. There's no experiment here, just the grassy backyard of an Italian villa near a pleasantly burbling fountain. The terra-cotta walls and wooden roof beams are done in stunning detail, with cracks and smoke stains ren-dered in painstaking variety. A fire in the hearth pops satisfyingly; the trees sway in a gentle wind, which is carrying updrafts of white dandelion fluff and enormous blue butterflies.

I am struck by how comfortable it is to be here. I want to peer through windows, walk through open doors, expecting the world to unfold endlessly in front of me. I start looking for the simulation's flaws, signs that this isn't a real place. It is perhaps too pretty. Its extreme rusticity gives it a stage-y theme park quality, the bright colors reminiscent of plywood and paint. I find myself wishing they had rendered, say, a parking garage, something with the dinginess of ordinary life. And there is the slight tug on the back of my head as Karutz, holding my helmet cables, tries to prevent me from walking into real-world walls as I heedlessly ramble through the imaginary landscape. But mostly I think about how pleasant it would be to roam here in an adventure game, or watch a dramatic story unfold at the villa, or just rest in the garden. I drop down on all fours to peer at the blades of grass beneath my feet, won-dering if they will pixelate up close, spoiling the illusion. But they don't. I pad my way over to the fountain and gaze into the water, which is sparkling and rippling in the afternoon breeze. And this is when the world's one true off-note hits me: I have no reflection.

Sometimes the moment the spell breaks is the moment that proves how tightly it's bound you. Because while I am internally debating how much the virtual water looks like real water, this is how things look to Karutz, patiently holding my tether in the dark, empty studio. I am so caught up in the dream that I am on my knees, head cast down, arms outstretched, searching for myself in a fountain that isn't there.

THERE IS ONE OTHER QUESTION the folks in Bailenson's lab like to ask, and it's a big one: In virtual worlds, do we have to be *us*? The lab started out with subtle changes to the human appearance: making your avatar prettier, taller, fatter, or older than the real you, even changing your race. (In psychology, this is usually called "perspective taking," or envisioning yourself as another person or in an altered state.) But in the virtual world, there's no reason to stick with your human form. In fact, learning how to occupy another body might do us some good. "We've known from our past research, when you occupy a human avatar that you gain empathy towards that human. So you can reduce racism, ageism, sexism," Bailenson says. "Does the same thing work with a nonhuman?"

Nine months later, I head back to the lab to find out. Once again, it is dark and quiet and Cody Karutz is cheerfully strapping devices to my appendages. He fits me with fabric kneepads, then infrared markers around my wrists. He helps me into a nylon pinnie—the kind kids wear for soccer scrimmages— and sticks two infrared markers along my spine. On goes the helmet. Then he asks me to get down on all fours and he flips the simulation on.

I am a cow.

I am a cow in a lovely pasture, a green expanse ringed by distant barns. In front of me is another cow—Karutz explains that this is a mirror image of my avatar, put there because, given the physics of wearing a helmet while crawling, it's hard to look down and see my own body as a cow's. I make an involuntary "Aww!" noise because my cow self is adorable, a pint-sized brown-and-white calf with tiny curved horns and a fat body on spindly legs. I lift my right hoof and my cow double does the same. I trot around a bit, getting the layout of the field, watching as my cow self does, too. VR folks use the term "body transfer" to describe shifting your consciousness to an external representation, and as the cow copies my movements, that's just what we're doing.

"Welcome to the Stanford University cow pasture," booms a voice overhead. "You are a shorthorn breed of cattle. You are a dual-purpose breed, suitable for both dairy and beef production." It is momentarily jarring, being told I am suitable for beef production. But I roll with it, as the voice gives me instructions: Walk over to a feed cart and eat. I do my best to position myself over some hay. My cow double does the same. The voice ticks off some mind-blowing stats about how much weight we must gain—three pounds a day— in order to bulk up to 600 pounds. I wonder whether I should mime chewing, which seems oddly natural, even though no one has requested it.

Now the voice tells me to walk to a water trough. My cow double and I loop to the left. I can see a cattle prod hovering in midair. In the real world, it's a wooden dowel with an infrared marker attached to the end, held by a lab assistant. It's not working today (this is a very early version of the study) but normally he would have jabbed me with it lightly. I would have seen the cattle prod coming at me while I felt it press into my sides. This is "synchronous touch," Karutz later explains, another way to produce body transfer. Today, it's just floating nearby, so, unprodded, I stand over the water trough as the voice tells me I'll drink up to 30 gallons a day.

"Please turn to your left until you see the fence where you started," says the voice. "You have been here for 200 days and reached your target weight. So it is time for you to go to the slaughterhouse."

I was not expecting this. A wave of sadness and horror hits me with the word "slaughterhouse." The suddenness of the announcement, the feeling of being trapped, the guilt and responsibility I feel for my cow avatar, who I somehow feel *is* me, but who I simultaneously feel is younger and more innocent and who is, I should point out, a *vegetarian*—it's remarkably heavy for having been in this virtual life only a few minutes. The part of me that is a cow dutifully walks toward the fence. The part of me that is a person is yelling. It's unbidden, startling even me, an anger borne of nervousness. "That is brutal!" I shout at no one in particular.

The simulation presses on, telling me to face my cow avatar. She gazes back at me, as innocently adorable as ever. "Here you will await the slaughterhouse truck," says the voice. The floor begins to vibrate. I hear the grinding of approaching tires and the beep of a truck backing up. I feel a rush of real fear as the world shakes noisily around me. I scan my head from side to side, wondering where the truck is going to appear. What will happen then? But it doesn't. The experiment is over. "My God, you guys," I hear myself muttering in relief as Karutz unwinches me from the helmet.

Had I been in the real study, the follow-up would have gauged whether I now felt more empathy for cows, and more broadly, my feelings about animal rights. And the cow—cuddly, familiar, a fellow mammal—is just the beginning of where the lab is headed with this idea. Moments before I came in, Karutz had been making tweaks to an experiment in which participants will become a coral stem in a coastal reef, an even more unfamiliar body configuration. The coral is immobile, a brilliant purple, with branches that just vaguely recall arms, the only nod to the human form this simulation makes. It's situated in clear blue water, surrounded by a passing jumble of sea life. For synchronous touch, a fishing net will bump into you while a lab assistant pokes you in the chest with the dowel. This time, you will listen to facts about ocean acidification, caused by water absorbing the carbon dioxide produced by burning fossil fuels, as you watch the sea around you slowly die—first the sea urchins, then the fish that feed on the urchins, then the sea snails whose shells become corroded by the acidifying waters. As the animals disappear, the water becomes grayer and the rocks coated in algae. If you look down you will see your own body withering, until a chunk falls off onto the ocean floor.

"So in both of these you have to watch yourself die or almost die," I point out flatly.

"There are some very dramatic effects," agrees Karutz.

The lab is working with marine biologists to design the coral scenario, which subjects will experience as a VR environment, a video, or simply audio. Ultimately, says Bailenson, they're testing how well viscerally experiencing the dying coral works as an educational method. "Do you learn better, do you care, are you more motivated to learn when you become the coral?" he asks. They'll also track an empathy-related measure—maybe subjects' willingness to donate money or sign petitions for ocean-related causes.

There is also a more technical cognitive question behind all of this shape-shifting: the idea of homuncular flexibility. The homunculus, or "little man," is a way of visualizing how the cortex maps senses and movements onto the body. The areas that innervate different parts of the limbs, trunk, and head appear in this cortical strip in roughly toe-to-head order. But because the face and fingers are so sensitive and dexterous, they're more densely innervated and take up more space. If you drew the body of a "little man" based on this cortical schematic, he would have two fat lips and clown hands.

In VR, says Bailenson, "Homuncular flexibility asks, if you put someone in a body that is decidedly nonhuman, can they learn to operate it?" So imagine, he says, as Jaron Lanier—his friend and mentor—had attempted during early experiments with novel avatars, "you put someone in a lobster. A lobster has eight arms. Moving the first arms of the lobster is very simple. You move the physical two arms, and the virtual arms do what the first two do. How do you move arms three through eight?"

That's an extraordinarily magical question and also an extremely practical one. Think of using your avatar to manipulate digital objects, says Bailenson, like in the film *Minority Report*, based on the Philip K. Dick story, in which Tom Cruise plays a futuristic police officer. "You remember that scene where Tom Cruise is playing with all that data and using his arms?" asks Bailenson. "*Why is he just using two arms?* The data is all digital. If people can learn to control eight arms, then they'd be more efficient." Or, he says, think about using virtual environments to manipulate real-world machines.

You could have "many to one" control, in which a group of users operate a "team body," or "one to many," in which one expert controls multiple devices. Consider the military, he says: "The single best plane operator, why is she only operating *one* plane?" Or what about telerobotics? Think back to Dr. Wren and the da Vinci. Wren has two arms. The robot has four, driving three instruments plus a camera. Right now Wren can swap between two instruments, or ask an assistant to take over the third, but has no way of moving all at once.

So this is what we're going to try on my last day in the lab. Andrea Stevenson Won, a lab doctoral student, has built a scenario that will give me a virtual third limb, and we'll see how quickly I can learn to control it. Karutz fits me with the helmet and straps infrared markers and small plastic accelerometers to my wrists. Then the lights go out, and I'm staring into a virtual mirror at my avatar, a silver outline of a body. It's got the two usual arms, which I can operate by moving my own normally. But there's also this enormously long armlike protrusion sticking out of my chest. This limb has no elbow joint, and just the barest hint of fingers. Karutz gives me a few seconds with the mirror to adjust to my new limb, which is controlled by rotating my wrists. One of them—he doesn't say which—controls horizontal movements, and the other its lateral movements. I hold my hands out stiffly in front of me and jiggle my wrists. My third arm flips back and forth like a windshield wiper. This is all the training I get.

Then the mirror blinks off, and I'm looking at cubes hovering in space, just close enough for me to touch with my own fingertips. On my left, there are nine blue cubes; to the right, nine red ones. Every now and then, a cube will turn white. When it does, I must touch it with my real hands.

A couple of feet behind these are nine green cubes. When a white cube appears here, I'll have to touch it, too. "They are too far, so you can't use your normal limbs," says Karutz. "So that's why you have a third arm."

OK. Ready. A cube lights up in the blue set. I tap it with my hand. The cube flashes and emits a delightful shimmery zing before turning blue again. *Easy.*

Now one back in the green array lights up. I'm mentally prepared for this to be tough. Karutz is, too. He's about to offer some encouragement when I just reach out with my third arm and . . . touch it.

I have no idea what I did. I just did it. "*Nice!*" Karutz says.

I make an awestruck noise and keep going. The cubes flash, and I smack them. Real arms, fake arm—it's weirdly natural. The mental and muscular math required to make the third arm move happens subconsciously; somehow my two real wrists direct the imaginary one. I'm hoping this will be an actual job skill in the future. *Ask me about my homuncular flexibility!*

In fact, Stevenson Won concludes that people can adapt within five minutes, and that subjects given a third arm did better than a control group that only used their real arms and stepped forward to tap the green cubes. In an earlier phase of the study, she also found that people readily adapt to having their arms and legs switched, or to having their legs be able to reach extra far. I'd tried this the previous fall, and found it easy to complete the task—popping virtual balloons floating in midair—while using my arms to control my virtual feet, or kicking over my head with my suddenly superflexible legs. I mean, it hadn't been pretty. I'd lumbered around the room swinging my limbs like a deranged robot. But it got the balloons popped. The idea, she says, was to see if people would switch to using their feet, rather than their hands, which were better balloon-popping tools in both conditions. (They did, although they performed the task better when their legs were given extended range.)

Thinking about tool use is important in homuncular flexibility research, says Stevenson Won, because there are parallels between how we learn to use them and how we learn to control novel bodies. "People are very good at quickly learning to use new tools, and tools can be considered as extensions of the body," she says. In the third arm study, subjects are put into one of four conditions. Some see a limb attached to their chest, as I did. Others see it floating near their body, or as a sort of metal cylinder protruding from their chest, or as a hexagonal shape floating beside them. In other words, it can look like either a tool or a body part, and it can be attached to you or not. And that might change how you learn to use it. As Bailenson puts it: "Is it a hammer or is it your arm?"

There is a way to test this, too. After I have tapped my way through the cube task, Karutz makes the cubes vanish, and now a bull's-eye hangs in midair. He asks me to place my third arm at its center. I do, and there is a roaring noise and a bright light. My brain registers this as: *My hand is on fire.*

I yell. My shoulders and neck involuntarily go into a deep cringe. And that is exactly what the lab wants to know—how I react to a threat to my imaginary limb. "If it's a tool, if somebody sets it on fire, you shouldn't flinch," says Bailenson. "And if it's your arm, you should."

And look, by this point I have spent a lot of time in this lab. I have read a ton of their papers and interviewed them relentlessly on their methods. I can see how the trick is done. I know it's a virtual hand and a virtual fire. But that little tiny freak-out moment?

It's real.

TEN

· ·

Augmented Reality

ROB SPENCE'S EYE is in the shop.

Spence, better known as Eyeborg, is in his living room smoking cigarettes in front of the space heater. Outside the wind is twisting orange maple leaves and the Toronto winter's first snowflakes into delicate rivulets. His appearance is a bit piratical: His dark hair is tousled, he's sporting a few days' worth of stubble, and over his right eye is a patch. It covers an open socket punctuated by a motility coupling post, onto which he can snap a prosthetic eye. Sometimes he wears one made of glass. But other times, he snaps in a camera.

A laptop on the coffee table is open to the designs of his cameras, an upcoming model and previous ones. The first working model looked more like a mysterious robot part than an eye, an oddly rectangular assembly of battery, transmitter, and camera bits. There was another version that could, at certain moments, illuminate a red LED. "It's *Terminator*," Spence says wryly. "When you have a missing part of your body, and you read a lot of comics, why not?"

Yet the next iteration is being built to look exactly like Spence's real eye. He pulls up some photos of its shell, hand painted by his ocularist, who re-created his pale green iris and used thread to mimic the eye's tiny red veins. The camera will be behind it, its lens peering through the pupil. Once the guts of the new model are updated, Spence will use it for his work as a documentarian. "For me, it's a toy," he says. "A really, really cool toy for a one-eyed filmmaker."

These cameras don't connect to Spence's brain, or communicate to his retina, as Dean Lloyd's implant does. They are wearable devices that sit on the border between internal and external, transmitting wirelessly to a remote receiver. Spence uses one that's essentially a baby monitor, a white plastic box with a video screen. But in theory, the feed can be beamed anywhere—someone else's laptop, the entire Internet.

Spence's eyecam, which debuted in 2009, is a sort of proto-augmented reality device, one step toward wearable sensory enhancement. Augmented reality (or AR) devices are usually worn or carried, not permanent, endowing the wearer with perceptual modifications or conveying otherwise hidden information to them. Some consider them a half-step between virtual reality and implantable devices that would directly affect the brain; the technological middle ground between a room's worth of rigging and a chip buried in flesh. Augmented reality's signature is an overlapping of the physical and virtual worlds; indeed, the field is sometimes called "mixed reality."

AR is still in its infancy, so the rules of the game, and the form factors of devices, are just emerging. The products creeping onto the market are mostly accessories: watches, rings, garments, cell phone apps, and glasses, which put a camera *over* the eye, rather than *in* it. While there's not universal agreement about what constitutes an "augment," the simplest level "is basically overlaying graphics on the real world," says Ori Inbar, cofounder and CEO of the not-for-profit industry organization AugmentedReality.org. He also organizes the annual Augmented World Expo (AWE), an energetic figure who can be spotted roaming the crowd in his signature pink and orange jacket, through which a light display flashes: "I'm in AWE."

Overlaying graphics and text is essentially what Google Glass did; the technology giant rolled out these sleek spectacles in 2013 to a limited number of "Explorers" before stopping sales in early 2015 in anticipation of a design reboot. The original design displayed functions like mapping and email using a miniprojector in the arms of the glasses that beamed light into a prism that jutted across the right eye. Glass offered a camera as well as the ability to do voice-activated Internet searches and phone calling. While the company's high visibility made Glass AR's de facto flagship device, and the center of early public debate over wearables, that's a position it held uneasily. In its online FAQ, Google declared that Glass is not immersive, "certainly not augmented reality" and that "the screen is inactive by default." (That raises the question of what Glass is when the screen *is* active, but Google's press team did not respond to interview requests.) Still, while many insiders give Glass props for piquing public interest, they say it isn't true AR, more like putting a smartphone over the eye. A product like Glass provides information in context, Inbar says, but because it only covers a corner of the wearer's field of view, it doesn't truly paint a second world over the real one.

Many nonwearable AR devices require pointing your cell phone or tablet at a physical object to see something virtual painted atop it: Hold it over an ad or a magazine to see the model come to life; point it at the sky for information about stars or the planes flying by; point it at a biology textbook to learn how the heart works. One of my favorite ideas, on display one year at AWE, was built by students at Switzerland's École Polytechnique Fédérale de Lausanne, who gently helped me press a temporary tattoo of a looking glass onto my arm. Then they led me to a special mirror, in which the tattoo seemed to animate. An eyeball formed at its center and wept black tears until the looking glass filled with them and cracked into pieces. But as charming as these superimpositions can be, other developers say they are going further, creating multisensory devices that will truly alter perception—that will "program reality," as Inbar says. He argues that AR is different from VR, his former field, because rather than being immersive, it is deliberately semitransparent. Eyewear is see-through, instead of occluded like VR displays, so you focus beyond the screen. Portable devices let you move freely, using your hands and senses as usual. "The idea is to actually be in the world, in the here and now, but to enhance it," Inbar says.

In fact, Inbar was drawn to the field because he felt that his own children were too comfortable in front of screens. Originally, he thought most applications would be kid-centered ways to "gamify" learning or chores by making objects appear interactive, or letting you rack up points for completing tasks as you moved around in the real world. But he soon realized the earliest adopters of AR glasses were likely to be in workplaces where people need to get information while keeping their hands free. So, for example, smart glasses could give the illusion of X-ray vision, letting a doctor see scan results superimposed on your skin, or an electrician envision wires within walls. Others can guide people through complex tasks, everything from warehouse packing to engine repair. Remember the scene from *The Matrix*, Inbar asks, when Trinity instantly learns to fly a helicopter using instructions ported into her brain? "Augmented reality can do just that, minus the plug," he says.

Sci-fi and cyborg references are rife in the AR industry, and why not? They've long been our model for augments, particularly eyewear, everything from Geordi LaForge's VISOR on *Star Trek*, which let him "see" the entire electromagnetic spectrum, to *Neuromancer*'s Molly Millions, whose permanent mirrored lenses gave her night vision and a digital readout. Eyeborg's camera isn't that advanced, but it does give him some superpowers. He can take pictures with his eye. His ability to record gives him total recall—for two hours at a pop. He can connect his vision to computers and the Internet. He likes to point out that if you could combine his eyecam with retinal implant technology and slap on some graphics, you would finally get "Terminator vision"— real-time sight supplemented by tactical information. That said, he's only got part of the puzzle. Like Lloyd, Spence is a Model T: an early version with rudimentary powers.

Spence damaged his eye when he was nine years old and messing around with his grandfather's shotgun, holding it so the sight recoiled into his eye. His family rushed him to a hospital for surgery, "and they temporarily saved the eye, but it was legally blind after that," he says. "I had the same kind of vision you would have if you were looking through a Coke bottle." As an adult, Spence became a filmmaker. His first documentary was a satirical ode to his hometown called "Let's All Hate Toronto," in which Spence appears as what he calls one of his "metacharacters," Mister Toronto, its eyepatch-wearing ambassador. His vision deteriorated during the filming process, so in 2007 his doctors removed what remained of the eye. That's when he came up with the camera idea. Eyeborg, his second metacharacter, emerged during its construction.

People always ask Spence why he decided to build an eye camera, but it seemed obvious to him. He's been a sci-fi fan since forever. He's got 3,000 comic books in a closet somewhere. On the mantelpiece sits his beloved Six Million Dollar Man action figure, the kind with the plastic inset that lets you look through the back of his head so you can "see" through his bionic eye. (In the TV series, the eye had a zoom lens and could sense heat and see in the dark.) If you're *that* kind of guy, people naturally suggest *that* kind of thing. "It's not original at all," says Spence. "It's just that *I did it*."

Or rather, he started telling people he would. Without a budget, he appealed to the geek pride of engineers who might help out in exchange for, mostly, bragging rights. He cold-called electronics companies, attended conferences for camera makers and hackers, and talked to an eager tech press, selling the sizzle well before the steak. "You can get a lot done with sizzle," he says.

Although he's worked with many partners, most of his help came from electronics companies OmniVision and Rf-Links.com, as well as engineer Kosta Grammatis, who moved into his house and knocked out a prototype in two months, and Martin Ling, an electrical engineer in Scotland who, at that moment, was working on version four. Over time, the pace of the project ebbed and flowed, but this homebrew approach allowed them to circumvent considerations that slow corporate or academic projects—they didn't have to contend with government agency approval, university human subjects testing regulations, grant writing, or creating a mouse model. They just did it. "It's kind of like I'm Steve Jobs, but I don't have any money and I'm not that smart," says Spence. "But I do have a great team, and we do innovate and we're sort of nimble."

They had some tricky challenges to solve, Grammatis recalls: "Batteries blow up. They're dangerous. The environment the eye is in is really humid.

All the materials have to be biocompatible so we don't hurt Rob. It has to have a certain amount of battery life that makes it usable. The frequencies that the radio operates on have to work well through human skin." But once they were ready to showcase their work, they were deluged with attention from the press and the hacker world, which embraced Spence's cyborg nature. Back in 2009, Grammatis points out, filming from the perspective of the eye was fairly novel. This was long before Glass or the GoPro, a small, lightweight camera now commonly used for shooting first-person action scenes. Even the iPhone had only been around for two years. "No one had ever seen video from that perspective," recalls Grammatis. "We were getting invited to speak all over the world. It was a wild ride. We won *Time* magazine's best invention in 2009. We were in *Ripley's Believe It or Not*. It was crazy!"

Spence is not the first person to hang out his shingle as a semiprofessional cyborg, not even the first Torontonian with some kind of eyecam. That honor goes to University of Toronto engineering professor Steve Mann, who has been continuously wearing over-eye cameras of his own construction since 1978. He calls his the EyeTap, and to his resulting perception "augmediated reality." (Mann was an early collaborator on Spence's project, but they parted ways.) Among other notables: Since 2004, British artist Neil Harbisson, born color-blind, has been wearing what he calls his "eyeborg," an antenna that arcs over the top of his head and translates color into sound frequencies using a chip that was implanted on his occipital bone. He later extended its range to convert nonvisible infrared and ultraviolet wavelengths into sound. In 2010, New York University arts professor Wafaa Bilal had a camera he called the "3rdi" implanted in his scalp, a literal "eye in the back of his head," which transmitted images to a website. And since 1993, wearable computing expert Thad Starner of the Georgia Institute of Technology, who helped develop Google's Project Glass, has been wearing an evolving system dubbed "The Lizzy." (The name is a tip of the hat to the Model T, once called a "Tin Lizzie.") The original model was mostly worn around the waist, but included a head-mounted display and a keyboard attached to the hand.

And there is an entire army of people wearing early neural and motor prosthetics, many of whom Spence filmed for a 2011 minidocumentary to launch the video game *Deus Ex: Human Revolution*. The game stars cyborg hero Adam Jensen, who sports an eye camera and augmented hands. In his film, Spence introduces himself as a fellow cyborg, and travels the world interviewing others, including Finnish man Miikka Terho, a clinical tester of the Retina Implant AG device, and prosthetic arm wearer Jason Henderson, a gleeful exchange in which Spence exclaims, "I am now filming your bionic hand with my bionic eye!"

Spence's camera has its limitations and oddities. It's only got two hours of recording time. The video is in color but the resolution's not nearly as good as natural vision—he compares it to cell phone video from the year 2000. It's hard to broadcast from inside a socket, which, Spence says, is "like sticking a transmitter inside a ham." Sometimes when he's not getting good reception, Spence says he whacks himself on the side of the head to clear it up. But the eyecam is a fascinating mix of human and machine. "It's a part of his body, so it's a very intimate perspective to view the world from. You see what he sees, exactly how he sees it," Grammatis says. Because Spence has a working eyelid, he blinks, so every few seconds the screen goes black. His lashes flick in and out of frame. And unlike a camera mounted on a tripod or shoulder, it follows his gaze; the coupling post attaches to the muscles in his eye, which still move normally. This has given him some wonderfully novel filmic possibilities, and some problematic ones, like the time when a British TV presenter, watching his feed on the monitor, joked, "You're looking at my legs," and a blushing Spence fumbled something about always being careful to look women in the face.

As a filmmaker, Spence likes these irregularities, a step away from what he calls "third person eye of God" storytelling, in which the effects of dollies and rigging make it appear like "magic angels" are moving the cameras. "Mine is much closer to a first-person stream of consciousness, because it's more awkward; there's less rules; there's a lot of glancing and blinking," he says. He considers himself more aligned with the "shakycam" style popularized by *The Blair Witch Project*, a low-budget horror film purportedly shot on handheld cameras by lost students running through the woods, which amplified the terror through the conceit that the results were not a movie at all, but "found footage." (It was also famous for, at least apocryphally, making the audience motion sick.)

So Spence avoids making lofty claims about his camera's capabilities. ("The power it gives me is the ability to get press," he deadpans.) But it made for an excellent precursor to the arrival of commercial AR products, because it showed how intimately a person could wear a sensory device. It publicly raised technical, artistic, and ethical questions posed by wearables that collect images or other data. And most of all, it was a working proof of concept from the bleeding edge of reality research, which is a mix of corporate and hacker science, one foot in the lab and the other in the garage. Spence built his device with volunteer labor and strained credit cards, not big research dollars. It doesn't have awesome graphic overlays or a more capable camera because they couldn't afford them, not because they're impossible to build. But it's not hard to imagine a commercial company trying to give wearers sensory superpowers—zoom, facial recognition, contrast enhancement, even night vision.

In fact, somebody already is.

ARTHUR ZHANG IS HOLDING A TINY GLASS VIAL. Floating inside is a contact lens, its interior component as shiny and metallic as a dime. "The silver color that you're seeing is all the light that's being reflected at you," says Zhang. "It's like a mirror."

Zhang is a senior member of the technical staff at a company called Innovega, and the augmented reality system they are developing is the iOptik. You wear the metallic lenses as well as special glasses, which beam images toward the eye, creating a virtual overlay on the real world. Their design stands out in the AR world; a contact lens touches the body more intimately than glasses. It's their solution to one of the biggest technical challenges confronting the field: allowing you to clearly see far away *and* a computer display near your eye at the same time. That's crucial, they think, for an AR experience that's convincing yet not too distracting.

People who have tried to solve this problem with glasses alone have had to produce larger optical systems to create a larger field of view—but that means clunkier glasses, which won't fly with consumers. So instead of making the eyewear accommodate the limits of human vision, says Zhang, "let's change up human vision." The contact acts as a dual gateway, letting ambient light—required for distance vision—through normally, but re-aligning light coming from the display, so both are simultaneously focused on the retina. "It provides us superhuman focusing ability," says Zhang, essentially turning your eye into a magnifying glass, so a small display will seem large and clear.

There are several iOptik eyewear models in the works. A glanceable version looks like a pair of orange-tinted beach sunglasses with a display screen near your right temple, which you must shift your eye to see. A transparent version looks more like ballistic eyewear; in this version, a projection system reflects images to the eye, so you can see them directly ahead. Zhang thinks they can provide a 40-degree field of view for the glanceable version and 90 for the transparent one, about what's needed for an immersive experience. (For reference, Glass, considered highly transparent, has an about 15-degree field of view, and the Oculus Rift, a VR display with no transparency, is about 100 degrees.) Two other models—one for sports training, one for mobile entertainment—will look like sunglasses, and will hook into a cell phone or tablet device, then stream video to both eyes. The athletic glasses would also display real-time performance metrics; the entertainment version would offer a nearly 70-degree field of view, which the company compares to watching an IMAX film.

Innovega, a development company, plans to license out the hardware and lens manufacturing. You'd get a scrip for the lenses through an optometrist, who can add vision correction if you need it. (Contact lenses are medical devices and must be prescribed, even without vision correction.) Then you'd buy iOptik lenses at the pharmacy, just as you'd buy, say, Acuvue. If metallic eyes aren't your thing, Innovega is working on more naturally colored versions that will only slightly darken the iris. You could wear them all day, and put the AR spectacles on when needed—reading glasses for the cyborg future.

To see how iOptik works, we head into the back room of the company's San Diego lab space, where engineering director Jay Marsh has just finished up at a workbench covered in hand tools, calipers, and molding clay. He's set up a mannequin head with a camera peeking out of the eye. The "eye" camera is wearing one of the silver lenses. Over that, the head is sporting the glanceable glasses. The camera is hooked up to a projector, so we can see what the dummy head does: us, looking at it. Zhang plugs his cell phone into the whole apparatus, and now superimposed over the right half of the head's view is a map of our location. "That's basically a desktop in front of your eye," he says.

Like many AR designers, Innovega only builds the platform, and will rely on others to develop applications for it. Early AR apps have been graphically simple, often repurposed phone apps—email, weather, and the like—although specialized ones are on the way. These, combined with more immersive transparent displays, eye tracking, and a camera, could eventually offer what Zhang calls "pixel-perfect registration," a true mix of real and fake. Imagine that you're playing laser tag with a friend, but you see them as their avatar. Or instead of using a map, a line appears on the street to lead you to your destination. Maybe the flat surfaces around you become screens for personalized advertisements, à la *Minority Report*.

AR could also give you super vision: Innovega has worked with the military on projects exploring how to adjust contrast levels for transitioning between bright and dim conditions—imagine a soldier in a desert environment peering through a darkened doorway—as well as using a camera sensor to detect muzzle flashes. They've also considered creating a magnifying version for users with poor eyesight, and collaborated with a lab at UC San Diego that created a contact lens that lets users zoom in and out by winking. And while they're not developing it now, Zhang points out that if you added an infrared camera, you could give someone night vision.

In the near term, they see the most applications for industry, where even simple graphics and text can be helpful. Marsh, who worked on an auto assembly line during his engineering training, points out how useful it would be to receive instructions as you work. As the car rolls down the line, you'd see an overlay of where, say, the wiring goes. "You could be told it has to snake through these different holes in this way and clip in this place—and oh, by the way, you missed a clip," he says.

Indeed, one the first companies to actually bring AR glasses to market was New York–based Vuzix, which rolled out its M100 Smart Glasses in late 2013, largely to a workplace clientele. A few months after my visit to the Innovega lab, Dan Cui, vice president of sales and business development for Vuzix, gives me a remote demonstration of how workers are using the M100. He puts on his own pair, connects them to his computer, which is videoconferencing with mine 3,000 miles away, and then suddenly I can see through his eyes.

Next, he holds up a photo of a packing box and looks at its bar code, much as a warehouse worker would do for an actual box. "The computer on the headset is going to see the box, it's going to recognize it as having something there that I need to understand, and it's going to pull up that information for me," says Cui. A graphic of a blue polo shirt pops up on the screen, with the numbers 16 and 30 over it. Next to that is a yellow square with the number 3 and a blue one that says "Store 12." These are picking instructions, explains Cui: "That says there are 30 shirts in there, they're all blue, they are size 16, I need to take 3 of them, and I need to ship them to store 12."

Then he shows me a few of the 100,000 apps that run on their Android-based device so far, all developed by other companies, many originally for cell phones. He holds up a white business card and says, "I can translate languages on the fly," as the lettering on it changes seamlessly from "Thanks, friend!" to "Gracias, amigo!" Next he opens a beta version of a facial recognition app, and looks at a picture of his own face. Up pops his name, job title, and company—all pulled from his LinkedIn profile. (Facial recognition has been controversial for privacy reasons; Google banned the development of such apps for Glass, and the companies developing them usually say that if they bring their services to market, people will be able to prevent their data from being used. In this case, Cui had opted into letting the app access his profile.) Then Cui shows me an app, designed for field service or emergency workers and law enforcement, that lets you watch the feed from eight sets of glasses at once; in this case, we're seeing images shot during a hurricane.

The M100 looks a lot like marksman's glasses, with a white plastic arm containing the camera and the display curving in front of the right eye. The company's next project is meant for two eyes rather than one, and they hope to overcome the need for increasingly large prisms (that field of view problem again) by using waveguide technology, which projects the light sideways through the lenses themselves, eliminating the need for a bar over the eye.

Because the field is so new, its players are all trying make their distinguishing mark, whether that might be waveguide technology or contact lenses or a holographic interface, the selling point for the Meta 1, made by Silicon Valley company Meta. (Steve Mann is its chief scientist.) These glasses track head and hand position so you can manipulate objects in the augmented world in front of you, a significant curiosity for an industry still trying to figure out how people should control their glasses. (Voice command? Gesture recognition? Cell phone interface?) At the 2014 Augmented World Expo, there was a half-hour line around their booth, screened off by a heavy blue curtain, and guests were invited in two by two. While the eventual consumer version will likely have a different form factor, this version, intended as a kit for developers to use, was boxy and had to be strapped on.

Engineer Raghav Sood fitted mine, telling me to hold my hands up. Through the glasses, I saw them as red outlines. A hologram of a blue orb appeared. As I followed it with my head, the glasses tracked my position. If I looked away, an arrow helped me find the orb again. Everything was touchable: If I poked an orb it burst, exploding into yellow flame. (This, I imagine, is what popping bubble wrap will be like in the future.) I could hold shapes in my hand, drag them across space, and release them someplace else. Sood showed me a diagram of a motorbike engine. "If you reach out to it, you're going to see the components flying apart," he said, as just that happened. "If you pull towards yourself slowly, it will assemble itself. Slowly you can see each individual screw slotting into place."

By 2015, the field had diversified even further, with AR and VR product demos stealing the limelight at that year's big tech conferences; not only AWE but the Consumer Electronics Show (CES), where the world's developers come to debut works in progress, and gaming conference E3. Among the year's early design standouts: At AWE, the Osterhout Design Group, which initially designed extremely rugged-looking AR eyewear for military and government agencies at around $5,000 a pop, unveiled an industrial model called the R-7 that would retail at about half the price, and included nifty accoutrements like a Bluetooth-enabled "wireless finger mouse" worn like a ring, and swappable lens coverings that attach magnetically, letting you switch between dark coverings for VR applications and more transparent ones for AR use. At CES, the Oculus Rift team revealed its newest prototype, the Oculus Crescent Bay, which at that point looked a bit like an electric pencil sharpener strapped to the face, but was widely praised for its hyperrealism. Shortly before E3, the company also debuted its ring-shaped Oculus Touch controllers. Users grip one in each fist, and the cameras and sensors inside each ring track hand and finger motion, letting users manipulate objects, pick them up, and drop them in the virtual world. At CES, a company called Avegant showcased its Glyph prototype, a portable movie-and-video-game theater, which is worn like a chunky pair of headphones when not in use. When you're ready to watch, you slide the headpiece over your eyes, where a micromirror array reflects images to them. And while none of these designs were exactly *subtle*, tech critics cautiously hailed the new crop as increasingly immersive, useful, and possibly even hip enough to wear. *Possibly.*

It's hard to predict how shoppers will react to these products once they're actually for sale, especially for gear meant to be worn publicly, not just during home gaming. Most designers say that less obtrusive, dorky form factors are the first step in courting non-geeks. So is coming up with a pleasing visual experience, so users don't get distracted or seasick. But the ultimate test will be convincing people that wearables are an improvement over the cell phone, that there's a good reason to put a computer on your head. Kayvan Mirza thinks that reason is speed. When I first met him in 2014, the CEO of French company Optinvent had been showing off their initial design, the ORA-1 spectacle, a few tables away from the Meta booth. Glasses are "ten seconds closer to your brain," he said. You don't have to fumble in your pocket every time you want answers, so they reduce the lag between decision making and action. "This is a bionic eye that you wear," he added energetically. "It's all about having more information quicker, faster, when you need it, always on."

Mirza is quite comfortable seeing AR as a predecessor to brain implants, which would modify perception even more rapidly and seamlessly. "It is kind of a hybrid half-step before we get to the true cyborg level of melding with the network, the cloud," he said. He pointed out that as computing technology has become smaller and more portable, it's also climbed higher up the body. Now it is finally reaching the head. "We've gone from back rooms to our desks to our laps to our pockets," he said with a wry smile. "It's kind of crawling up slowly to our brains, isn't it?"

RIGHT ABOUT HERE, we should probably talk about what a cyborg is. Sure, it's a commingling of human and technology. But where is the dividing line? Does wearing a pacemaker make you a cyborg? How about a watch? Contact lenses? Clothing? Is it some ratio of flesh to machine?

Manfred Clynes and psychiatrist Nathan Kline coined the term in 1960 in the context of space exploration, imagining how we might adapt to harsh new environments by creating "self-regulating man-machine systems," instead of protecting ourselves in Earthlike bubbles, because "the bubble all too easily bursts." These systems must be integrated with the body, they wrote, because "if man in space, in addition to flying his vehicle, must continuously be checking on things and making adjustments merely in order to keep himself alive, he becomes a slave to the machine." The purpose of the cyborg, they continued, "is to provide an organizational system in which such robot-like problems are taken care of automatically and unconsciously, leaving man free to explore, to create, to think, and to feel." Among the possibilities they imagined: osmotic pressure pumps to continuously deliver drugs, an "inverse fuel cell" lung replacement to remove carbon from the body, and shunts to re-circulate fluids. The cyborg, they wrote, would be a way to outrun evolution, modifying the body without altering heredity.

But the term has since become more encompassing. Feminist theorist and science philosopher Donna Haraway, in her highly influential 1985 essay "A Cyborg Manifesto," argued that we are *all* cyborgs, complex creatures not easily bound by dualisms like culture/nature, self/other, and reality/appearance. Writing during an epoch of dramatic miniaturization for consumer technology, she noted that devices were becoming ubiquitous, unnoticed, and portable. "Our best machines are made of sunshine; they are all light and clean because they are nothing but signals," she wrote. "Cyborgs are ether, quintessence." She saw the potentially troubling military and industrial overtones of technology moving onto the body, but found the cyborg concept liberating, a way of being proudly *mixed*: "A cyborg world might be about lived social and bodily realities in which people are not afraid of their joint kinship with animals and machines, not afraid of permanently partial identities and contradictory standpoints."

British cybernetics professor Kevin Warwick, who calls himself "the world's first cyborg," has been modifying his body via temporary implants since 1998. In 2002, an electrode array implanted in his arm allowed him to remotely control a robotic hand, pilot a wheelchair, and sense the distance of objects via ultrasonic sensors fitted into a hat that pinged signals to his implant. He also passed motor-related electrical signals between his own arm and his wife's, who was wearing a simpler electrode. In his autobiography, *I, Cyborg*, Warwick defines the term as "part-animal, part-machine, and whose capabilities are extended beyond normal limits," and focuses on technologies that link people to a system, so they "become nodes on a neural network." He sees this as extraordinarily powerful, because a cyborg can sense and communicate "anywhere the Internet can take them" and control objects beyond the reach of the physical body. "Your cyborg body extends as far as you have an electronic connection," he writes.

But not everyone thinks networking is a great idea. Adam Wood, cofounder of the group Stop the Cyborgs, uses a similar cybernetics definition, in which person and machine are part of a control feedback loop. To him, the device doesn't have to be implanted, or even constantly worn—what matters is how it influences behavior. "There's a positive view of us as cyborgs: The person is enhanced; the person is using a tool; they've got additional powers," Wood says as he sits in a London pub nursing a pint. "But equally, you can view it the other way round. You can say the person is controlled by the system."

Wood is one of the most vocal and articulate public critics of wearable technology, particularly Glass, which his group regards as a test case. Wood is no neo-Luddite; he works in machine learning. He started out thinking AR was pretty cool, but he's worried it's being too rapidly embraced for the sake of novelty, without adequate discussion of the ideologies and capabilities embedded within it. The networking aspect troubles him most; the more networked a technology is, he argues, the more cyborg it becomes. Wood likes to envision an X-Y diagram: One axis is how closely the technology is integrated with the body; the other is how much it's externally controlled. In the corner representing the extremes of both, you have *Star Trek*'s Borg hive mind. In its opposing corner, you have a hammer. Somewhere in the middle, you have most of today's AR.

A device that allows a person's data to be collected, position tracked, perception altered, and actions modified via the information fed to them poses some dystopian possibilities. Think about all the ways AR glasses might nudge behavior, Wood says. They might incentivize it: You earn points for accomplishing tasks, or get social media applause for sharing personal tidbits. Or the glasses overlay information atop everything you see, influencing your choices. Imagine, for example, walking toward this pub, and up pops a TripAdvisor review claiming that it's rubbish, so you move on. Or you search for pubs and get a curated list, but can't tell which options have been left out. "You're being controlled, because it's the system altering your perception," Wood says.

Choosing a bar may be a negligible decision, he says, but what if thanks to facial recognition these devices start giving us information about *each other*, sharing, say, our criminal backgrounds or dating histories? (It's not impossible: Court records are public, there is already a boyfriend-reviewing app, and the Web is awash in voluntarily shared personal details.) What if someone develops what Wood dubs "TripAdvisor for People," so we can rate each other on qualities like friendliness, trustworthiness, or—for business applications—our value as a customer? While such scores might appear objective because they're based on numbers, says Wood, we can't know the biases of the system behind them. What could make you a "bad" customer? Is it because you have a low credit rating (which might show that you're profligate with money), or because you spend conservatively at that store (which might show you're tight with it), or because you once called the automated help line and shouted at it (which might just show you're human)? And finally, what if the machines we increasingly trust to assist our decision making, filter our perception, and store our data are unreliable—prone to manipulation, hacking, eavesdropping, or erasure? In that case, he says, "in a very real sense, your extended self isn't fully under your control."

None of this is unique to AR. People have always incentivized behavior and socially pressured each other. We've delegated tasks to machines for generations, and we rate, locate, and "like" constantly via cell phones. But with a technology that literally comes between your sensory organs and the world, Wood says, "it can be done at a very fine-grain level." With such close integration to the body, "we're less likely to kind of notice any manipulation. It's going to be less obviously external," he says. "So it's not new in any fundamental sense. But it's *more so.*"

So far, the biggest tension between the public and AR has been over privacy issues, a legacy of first-generation devices being so vision-and-camera intensive. One of Stop the Cyborgs' first efforts was creating a flyer so cafés and bars could ban Glass or "surveillance devices" on their premises. Wearing a camera on the eye is different than, say, closed-circuit security cameras, Wood argues, which are mounted discreetly. "It's very personal. It's very visible," he says. Essentially, it makes people feel *watched*. The idea struck a nerve. In the San Francisco Bay Area, the epicenter of Glass, several establishments posted similar signage, and nationwide there were flare-ups between Glass-wearing customers and annoyed patrons or managers. Perhaps in reaction to these episodes, and to the rise of the epithet "glasshole," in early 2014 Google released an etiquette guide advising wearers to ask permission before photographing, turn Glass off when requested, and politely handle questions and stares. When the company halted sales the following year, tech critics chalked the reboot up to surveillance concerns and social awkwardness: Users felt weird with cameras on their faces; bystanders felt weird being within their eyeshot.

It's not the first time there's been unease, even violence, over on-body cameras. In an article for the BBC, Harbisson wrote that in 2011 police officers tried to pull his camera off because they thought he was filming a demonstration. The next year, Mann was attacked in a Parisian fast-food restaurant by three men, one of whom tried to rip the spectacles from Mann's head (they are bolted on) while another tore up the doctor's letter he carries explaining their function. In 2014, an Ohio man wearing Glass in a movie theater was questioned by representatives from the Motion Picture Association of America and the Department of Homeland Security, who suspected him of bootlegging. (He wasn't.)

But you can argue that AR gear is no more intrusive or unnatural than the omnipresent mobile phone—maybe even *less*. "People criticize eyewear like we're going to be cyborgs and glassholes and all this stuff," says Mirza, the Optinvent CEO, but he points out that we're already devoted to what he calls "some anonymous rectangle," which we constantly allow to distract us. "Come on, that's just rude!" says Mirza. "Excuse me, I'm going to pull out my cell phone right in front of you and stare at a four-inch piece of glass and start typing away with my fingers. *How human is that?*"

And some people honestly don't mind being photographed. When I was at Innovega, Zhang said he figures he's always being recorded. "As long as people are not misrepresenting me or stealing my credit card information and identity, I am OK with everything being public," he said. Um . . . *even if you're having a horrible hair day or you get caught picking your nose?* "If I have my finger up my nose, then so be it," Zhang said with a laugh. "We just need to figure out a compromise of how to protect those people who do want to be private and protected and those that are OK with it."

That said, Innovega's early designs don't have cameras, and they might not add them—"We may decide that the first product doesn't have a camera because we think that's a barrier to success," said Marsh. And when I caught up with him again in 2015, Mirza's ORA team was rolling out a beta version of their spectacles to developers, but was also working on an "AR headset" model geared at young buyers already comfortable with wearing bulky high-end headphones. The gleaming white pair revolving in a glass case at the AWE show had a moveable camera and display arm that users could pivot away when talking with people, an acknowledgment of how sensitive it was turning out to be to put a device close to the eye. "Consumers are still not ready to wear technology on their faces just yet," Mirza had concluded, but headphones, he continued, "have a 'cool factor' that smart glasses just don't have." They're more entertainment than tool, he figures, more hipster than cyborg. With headphones, he thinks, you'll get the benefits of "closer to the brain, always on" technology without the privacy issues raised by a camera parked in front of the eye. "Superpowers without looking like a dork," he promised. "That's what we're about."

Kurt Opsahl, staff attorney for the Electronic Frontier Foundation, a technology watchdog group that fights civil liberties abuses, points out that AR is still a niche market compared to smartphones, which collectively gather much more video, audio, and location tracking information. Even if AR devices become more popular, they'll still fall under state laws that govern recording equipment, including audio laws, which generally require the recorded person to consent, and video laws, which prohibit use in sensitive places, like dressing rooms. And of course, he points out, "the ability to have surreptitious cameras has been with us for a long, long time. If you go to spy supply stores, they will give you cameras that are designed to look like coat buttons or a pen." These have been used for good and ill—everything from investigative newsgathering to upskirt photos.

But there are key differences with AR glasses, he adds, like being able to film constantly. "You probably wouldn't walk around with your cell phone held out and with a camera pointed at people, but you might walk around with a Google Glass," he says. We also don't yet have norms about when and where use is OK, or to put it in a legal framework, how to handle the tension between your right to free speech, to express what you have witnessed, and someone else's right to privacy. Say you take your smart glasses to a party or private club: Should you film? Traditionally, Opsahl says, social norms have curbed invasive behavior; there are repercussions for being a gossip or a spoilsport. He expects the same pressures to apply here: "You might be able to take a photograph of somebody when they don't want to be photographed, but they might not invite you back to the party."

These are ideas Spence had to wrestle with early, as a documentarian with an in-eye camera. Spence is ambivalent about his high profile in the cyborg world; he became Eyeborg thanks to an accident, not because he had a profound interest in cyborg theory or privacy issues. But he now spends a lot of time publicly discussing cyborgs, and takes an expanded view of what it means to be one. "I think wearing a t-shirt makes you a cyborg, because it's technology and it augments a normal naked human body," says Spence. "There is a reason why we are taller and live longer and don't freeze our feet in the snow—shoes are technology." When he gives talks, he argues that "in the future, people will try to swap their bodies more. It's just a normal evolution from shoes to knee screws to contact lenses to laser eye surgery." And to forestall audience complaints that nobody is going to "change their body just for the hell of it, I put up a picture of Pamela Anderson," he says. "Because millions of people are already completely surger-izing, are cutting into their bodies, changing themselves, just to have nicer-looking boobs. It's common, it's fine, it's normal!"

He sees a double standard: We pay attention to novelties, but ignore older technologies that are actually pretty high-tech. Take people who wear glass prosthetic eyes. "Now, do any of those people get called cyborgs? No!" he says. "But put a 60-cent red LED light in there and wow, you're a cyborg." Still,

Spence likes to goose the system. His team dreamed up a "gag eye" with a laser pointer, so he could glance meaningfully at PowerPoint presentations—certainly another bit of nerd porn. And while he didn't set out to make think pieces about surveillance, it's something he's considered. "The way the camera works leads you to stuff," Spence says. "Like if you have a GoPro, you tend to make videos about guys jumping over canyons from the point of view of the helmet."

When we met in late 2013, Toronto was busy being shaken by the news that then mayor Rob Ford had admitted to using crack cocaine after a cell phone video of him appearing to smoke the drug was circulated to the media. "You may have heard of our mayor by now?" Spence asked drily. "So why is his life fucked? It's because of small, discreet cameras that he didn't know were recording. So people aren't really scared of Big Brother. They're more scared of Little Brother," a term made famous by writer Cory Doctorow, who used it as the title of a dystopian novel about omnipresent surveillance.

One of Spence's biggest influences is the idea of "sousveillance," a term coined by Mann meaning watching from "below" by citizens, as opposed to surveillance, the watching from "above" by authorities. While powerful, it's a double-edged sword. The police beating of Rodney King, caught on camera in 1991, sparked riots in Los Angeles and international outrage, one of the first major examples of a civilian filming an abuse of authority. Today, many communities are pushing for police officers to wear body cameras, hoping to improve their accountability to citizens. But also consider the 2013 bombing of the Boston Marathon, after which law enforcement asked citizens to share videos and photos that might help them catch a suspect. Some found it worrisome that authorities ended up with footage shot by citizens, not officers. (And even worse, amateurs parsing photos and police scanner feeds on the Internet made several false accusations.) Granted, the marathon was held in a public area, where bystanders have no expectation of privacy. But a person wearing an AR device can go into private spaces—like businesses or homes, where the Ford videos were shot—that the government cannot access without probable cause and warrants, and use it among people who don't expect their image to be published or turned over to police. "The issue with wearable devices is it makes every action and every space visible," Wood says. "If every space is visible, it can be measured and observed by whatever the dominant force and dominant ideology is."

Tiny cameras make surveillance easier for authorities, too, Opsahl notes, and more likely to be used. Back when undercover agents had to follow people around and set up discreet cameras in public spaces, surveillance was extremely laborious. "It was such high friction that the government would only do that under circumstances where they had a really good reason," says Opsahl. For the average citizen, "This created a bit of a privacy protection. That would be gone if everything that you could do, you now can do *all the time*, against *everybody*."

Wood also sees gray space in citizens using sousveillance to enforce *moral* codes, rather than the authorities enforcing legal ones. Imagine a restaurant patron bursting into a racist tirade. "It's not a criminal act, so it's of no interest to that type of [legal] apparatus," Wood says. "Whereas your individual at the next table may be offended by it, may record it, may get that person sacked from their job or something. So there is a 'moral enforcement of the mob' aspect to this kind of crowd surveillance." Maybe you think shaming racists sounds great, Wood adds, but an ambient crowd of cameras has no clear ideological polarity: You can imagine homophobes using wearables to out the patrons of gay bars just as easily as gay rights activists using them to film discriminatory behavior.

Does it matter if those cameras are on the eye, not in the hand? "It's funny, people feel like it's more potentially scary and damaging and an invasion of privacy," Spence says of his eye camera, although using other small cameras requires similar ethical decision making. Yet he agrees that it's hard to know when wearables are on; indeed, they may always be on. (Some use indicator lights, screen illumination, or voice commands to signal the taking of a photo, but not all do.) Moving a handheld camera to your face announces your intention to photograph, giving your subject the opportunity to smile or demur. When the camera *is* your eye, that negotiation is less clear.

For documentary purposes, this intrigues Spence. Maybe you can "shoot first, ask later," as candid camera shows do. Or maybe you ask up front, and a less obvious camera will help put your subject at ease. Spence has begun brainstorming his next film, *Let's All Hate Toronto 2*, which will focus on hockey (or more particularly, "why the Toronto Maple Leafs suck") and plans to use the eyecam as a "literal point-of-view camera" to do close-up interviews with players and fans. There's something special about face-to-face conversation, and Spence is sensitive to that delicacy—how can you harness it, without exploiting it? "To look somebody in the eye is considered one of the last truly human things you can do with another person," he says thoughtfully. "It's the window to the soul, not the window to YouTube, right?"

LET'S GO SOMEPLACE in the AR world without cameras.

Dr. Adrian David Cheok's lab workspace at City University London is mostly a desk. It's an old-fashioned behemoth carved in dark wood, covered with a marbleized green blotter, which gives the otherwise spare attic room a genteel Victorian charm. Computer science doctoral students Jordan Tewell and Marius Braun keep pulling magical things from within its drawers.

First comes a handful of rings, which cascade onto the tabletop. These are design iterations of a project called RingU, an experiment in long-distance touch. The prototypes are simple 3-D-printed plastic pieces with nobs where the jewel would be. Braun and I slide them onto our fingers. When he taps the button atop his, mine vibrates and the nob glows pink. I press mine, and now his shakes, its top gleaming orange. The message in this tiny hug is simple: I am thinking of you.

The RingU version likely to hit the market first will be superhero sized, capped by an opalescent plastic gemstone. It can modulate vibration intensity by how hard you press, and connects to your phone network via Bluetooth, "So you could be anywhere in the world and just send a hug to someone," Braun says. Using an app, you can change the color the gem glows, matching it to your emotional state. It's like a mood ring, except your partner can sense *your* mood, not their own.

Indeed, this whole lab is based on sharing moods and sending sweet gifts. While much of the AR industry is medical, military, or industrial, Cheok, a soft-spoken professor of pervasive computing with a Beatlesque haircut and a friendly demeanor, is interested in just making ordinary people feel good. "I think we can use these computer technologies to increase human happiness," he says. "That's what we want to aim for, increasing connection between families, between friends, between lovers."

By "pervasive," Cheok means technologies that are ambient in the environment. The ring is an offshoot of experiments at his former post directing Singapore's Mixed Reality Lab, where Cheok explored ways to send a hug through the Internet. In one study, designed to help people cuddle a distant pet, users would caress a chicken-shaped doll, and a real chicken wearing a special jacket would feel the touch. From that idea, he crafted the Huggy Pajama, so parents stuck at work could give their children a bedtime squeeze.

At the time, the goal was just to convey touch, but his lab soon began to explore how touch influences feelings. Combining touch with other sensory input—like video, scent, or colored lights—might make the feelings stronger, he thought, and even create some strange abilities. "What about creating a record of the hug of your grandmother?" he asks. "Even if your grandmother passed away, you could still not only see her video but feel her hug at the same time, or smell her smell." And the more he worked on touch projects, the more Cheok realized that nobody wants to wear a special garment or blocky gadget. Hence the ring: unobtrusive, socially acceptable. "Even if you were in a business meeting or in the airport," he says, "someone can still send you a squeeze to your finger and it's very unnoticeable and discreet."

When Cheok began his AR career, he focused on vision, as much of the field does. But while sight is good for communicating data, he felt, it's less apt for conveying *experience*, especially over the long distances that increasingly separate loved ones. Sight won't let you squeeze your boyfriend's hand or hug your grandma. "The next stage of Internet will be not just sharing information, but sharing our experience," Cheok says. "This will need all of the five senses, including touch, taste, and smell." These three, particularly smell, with its deep connection to the limbic system, are especially useful for conveying feelings. So unlike the hard sci-fi edge of many AR projects, Cheok's are often whimsical and nostalgic, built on the joys of everyday life: kissing, cooking, sending notes, smelling a familiar perfume.

Take Scentee, the next surprise Tewell and Braun pull from the desk. This is a small white capsule filled with a liquid scent, which you attach to your phone. If you and your friend both have the Scentee app, you can send puffs of scented air as you message each other, perhaps suggesting a café meet-up with coffee, or a date with roses. (Scentee went on the market in Japan in 2013; the cartridges only offer one scent at a time, although you can buy replacements.) The lab has tried advertising applications, too. In 2014, Oscar Mayer—here dubbing itself the Institute for the Advancement of Bacon—launched the "bacon alarm clock" campaign, sending 3,000 lucky applicants capsules that, when connected with a "Wake Up & Smell the Bacon" phone app, would rouse them from slumber with the sound and scent of frying pork.

Another project taps into Cheok's interest in bonding over shared meals. Working with Mugaritz, a high-end restaurant in Spain, his lab developed a teaser app to use before your visit. Mugaritz diners begin each meal by grinding ingredients for a savory broth to share. The app re-creates this experience by showing an overhead view of a mortar and pestle. If you twist your phone in a circle, the ingredients—sesame seeds, peppercorns, saffron—become powder as Scentee emits the broth's aroma, a pretaste of the meal to come and a way to recall it afterward. Because the broth making is a communal activity, Tewell says, from now on that scent memory will be connected to fond thoughts of your dining companions.

If you can hug through the Web, you can certainly kiss through it. (For the record, you can do more; there is an entire haptics research field called teledildonics devoted to remote sex.) The Cheok lab's newest tactile project is Kissenger—that's a portmanteau of "kiss" and "messenger," not a nod to Henry Kissinger—and conveys a kiss by phone. The challenge, says Cheok, is "we don't want to kiss someone through glass." The hard surface just feels weird. So in the first version, users exchanged kisses by smooching toy animals, plastic spheres with googly eyes, fuzzy ears, and paws that rendered them vaguely dog- or rabbit-like. The animals had giant pink lips, which relayed a vibration and the force of the kiss. The next version, which Braun extracts from a drawer, is a pair of lips that attaches to your phone. You lean into them as you videoconference with your kissing partner, and it uses force sensors and actuators to mimic the pressure of their lips, creating a more realistic kiss. "This will be much more neutral, probably," Tewell says, since it doesn't require you to lock lips with a doll. "Also intimate, because you will actually be really close to whoever else's face it is," adds Braun. Maybe they'll amplify the experience by using Scentee to release a perfume you associate with your loved one.

Working with touch, taste, and smell are, frankly, hard to do. That's why relatively few AR labs have tried it. While light and sound waves easily translate into binary code, you can't transmit pressure or chemicals via computer screen, so you need special hardware on both ends of the Internet connection. For RingU, both partners must wear rings; for Kissenger, they must own the lips; for Scentee, they must have their own cartridges. But there are a few notable explorers, many in Japan. A lab at Keio University invented "Tag-Candy," a lollipop inserted into a vibrating base that transmits touch and sound through the candy (and then from teeth to bone), creating the illusion of it fizzing like seltzer, exploding like fireworks, or even rumbling like a landing plane. One at Tokyo University created the "Meta Cookie," a head-mounted contraption that changes the flavor of a plain cookie to chocolate, almond, or maple by blowing scented air at the user's nose. One at Meiji University tried to alter tastes with electrified straws and utensils.

And a few months before I visited Cheok's lab, I stopped by Disney Research in Pittsburgh, Pennsylvania, where Ivan Poupyrev was exploring transmitting long-distance touch using air. Like Cheok, he is interested in AR that doesn't require elaborate gear. For a company that makes theme parks, and wants to create uniform experiences for hundreds of guests at a time, fitting gear creates difficulties: sizing, lines, staffing, awkward reactions. Plus, having to put something on "breaks the whole magic," Poupyrev says. It's better to work with what's already there, he says, and "air is the only medium which surrounds us."

His prototype device is called AIREAL, and it looks like a plastic box atop a camera tripod. Openings on five faces deliver vortices of air forced through a nozzle by a subwoofer. Each vortex is donut shaped, continuously drawing air up through its center, over the sides and back around. This gives the rolling air stability; it can travel more than a meter before dissipating. When it hits you, the low-pressure zone created by air in motion collapses and you feel a tap, anywhere from a hard kick to a soft flutter. Maybe you could try a different type of gas like helium, suggests Poupyrev, or change the temperature of the air, or put something inside the vortex—smoke, a scent, even fire.

In one of their early demos, they created the illusion of a butterfly landing on your hand. Two devices were mounted over a table, essentially a giant video screen. If you put your hand atop it, the image of the butterfly would dance on your palm, and AIREAL would pulse a stream of fast but weak vortices at it, carefully timed to the flapping of its wings. In other demonstrations, they created the illusion of birds circling your body, or of soccer balls flying out of a TV screen, which you could "block" with your hands like a goalie.

Touch feedback can be minimal and imperfect, Poupyrev says, because combined with video, it's enough for suspension of disbelief. Entertainment has different user demands than the kind of AR that, say, a plumber might need to fix a pipe. You don't want people to concentrate on the technology. When people visit an attraction, "we want them to feel as though they walked into the enchanted garden or enchanted house or they walked into the fairy tale," Poupyrev says. "That's the difference between full experiences and magic tricks."

Yet AIREAL may never secretly enhance your Disney experience. Only a few months later, Poupyrev had moved on to a new job at Google (he was leading Project Jacquard, creating textiles with conductive fibers woven into them, making them able to sense and transmit touch), and a Disney representative later said the AIREAL project was no longer under development. Still, even as a prototype, it had shown how charmingly AR could be rendered through a sense other than sight.

Cheok thinks the next step for augmented reality is to digitize touch, taste, and smell information, as we already have done for images and sound. "Before MP3, if you wanted to share a music file you'd have to pass someone a record or a tape, right? It's not very scalable," Cheok says. "After MP3, we could distribute music anywhere in the world. This is similar to smell and taste—right now it's very much analogue." Going digital means skipping the chemicals, and directly stimulating receptors. Cheok's lab has developed a small device that, when you hold it to your tongue, creates bitter and sour sensations by electrically stimulating taste buds. Next, working with a neuroscience lab at the University of Marseilles, Cheok wants to develop a version that can be used inside an fMRI scanner, so they can see how the neural reaction compares to chemically stimulating taste cells. He is also exploring whether you can directly stimulate the olfactory bulb, perhaps via an electromagnetic device worn on the roof of the mouth. While this research is extremely preliminary, Cheok thinks that ultimately you might use such devices to create experiences you can't have in the physical world, like a flavor that will "go from cinnamon to vanilla taste in one second."

So far, the limitations of first-generation smell, touch, and taste AR devices have largely exempted them from critical scrutiny; they don't involve cameras, track particularly sensitive information, or alter perception without you realizing it. But Cheok, who has written about the ethical dilemmas facing augmented reality, says this might not be the case if sensory devices become digital and implantable. He's been following the development of optogenetics in other fields, and foresees its applications for AR. "If you can connect optical fiber to neuron, that means effectively you've connected the computer world to the biological neuron world," Cheok says. "We may see within our lifetimes that we can directly stimulate, for example, a smell—not by stimulating the olfactory bulb but stimulating the brain centers, the neurons responsible for the perception." That changes the game, he says, because "if people wear these things all the time, there will be a real blurring between virtual reality and reality."

That said, both he and Poupyrev point out that people accept—even desire—that blur. Poupyrev notes that in many parts of life, we embrace artificial experiences because they nevertheless produce real feelings and memories. "If you go to Disneyland, is it reality?" he asks. "Or if you go to a movie, is it reality?" In fact, both say, we never fully experience reality, because our senses only pick up a narrow band of the world. "We can't see infrared. We can't hear the frequencies that bats hear or dogs hear," says Cheok. "So already our reality is a kind of modulated reality."

What's one more level of modulation?

AS AR MATURES, its community will have to figure out what they want from it, how to harness its benefits without unleashing the worst of its problems, and how to move beyond mapping, email, and other functions that replicate what you can already do on a cell phone. Inbar, the AWE conference organizer, likes to invoke the term "skeuomorphism," meaning using older conventions to shape newer media. Think of early movies, he says. Although they amazed audiences with how real they seemed—the story goes that the Lumière brothers' footage of a train steaming into a station caused panic in the theater—many borrowed the look of the stage, framing the action with curtains and using theatrical sets and props instead of venturing into natural settings. Similarly, Inbar says, "I think we have to invent the new language of interacting with the real world." If you compare AR to the film industry, he says, "we are probably in 1903 or so. We have the technology and have shown some really cool stuff—the train entering the station and so on. But we haven't invented the cut, the zoom, the pan, all these things that became critical to tell a story in film."

One of the most interesting cautionary notes comes from people who are considered disabled, who have already witnessed several generations of assistive technologies designed to augment the body. In this community, there is wide-ranging debate over how helpful and welcome these are. Dr. Gregor Wolbring, a disability, ability, and science and technology studies scholar at the University of Calgary, has frequently written about the social effects of new technologies, including prosthetics and implants. He argues that they create—and reinforce—expectations of what people must be able to do to be considered "productive" or "normal." He calls these "ability expectations," and points out that they change constantly. Until recently, nobody had to use phones, computers, or the Internet. These devices changed the world for everybody, but think about, for example, the special way the transition to remote communication affected lip readers and signers who once relied on face-to-face conversation. And with enough public buy-in, these skills became required. Livelihoods are tied to technology, Wolbring says; even something as simple as delivering pizzas requires a car and your ability to drive it. "So we already do this," Wolbring says. "We just use more technologies, and we never question what we want as a social structure, which ability expectations we see as important."

Even though early AR devices aren't particularly powerful, he says, we should think about which abilities we expect them to confer, who benefits from them, and who might get left out if they don't adapt. "The computer couldn't do much in the first generation, right?" Wolbring asks, but just as computer skills became mandatory for middle-class jobs, he expects AR to become part of work life. "For many of these things, there will be no choice. If you want to have a certain job, you will simply have to do or buy certain things," he says. Sure, you could refuse, just as you could refuse to use the Internet. "But you have to live with the social consequences," he says, which likely means damage to your income, your education, your circle of friends. "Now," he asks, "is that choice?"

As Wood might say, none of this is new—it's *more so*. Wolbring uses "ability creep" to describe the expectation that we will constantly have greater and more diverse abilities, an endless procession of "new normals." It's a creep that began long before electronics, no different than the sharpened flint or the harnessing of fire, any technology that conferred a survival advantage upon the holder, and forced everyone else around him to catch up or get left behind. The pro-tech will call it the soul of innovation; the wary might call it a rat race. It's an old cycle for an intelligent species, but it has a deeper resonance now that these new normals involve the body—and perhaps soon the brain.

One day as I sat with Kayvan Mirza at his AWE booth, he offered an unflinchingly Darwinian outlook on what might lie ahead. "It's kind of a scary thought, but you know what? If the guy next to me has a brain implant and he can do his job ten times quicker, and he's going to be more efficient, and he's going to see the world in an augmented way and have more information, that will make him, in a way, superior. I don't want to be left out," he said. Everyone will have to upgrade as much as they can. "It's survival of the fittest," he continued. "That's like self-evolution, right? We will take control of how we evolve."

If sensory augmentation devices begin to confer advantages, society will have to evolve, too. Almost everyone we've met in this chapter had the same answer for how to govern these new capabilities: social norms. As Opsahl pointed out, social change can keep pace with technical development; the courts can't. And as Cheok put it, you can only ban or ignore a device when just a few geeks are wearing it. Once your grandma's got one, he said, "we will just have to accept new norms of behavior." And that's a big question: What happens if AR makes it to mass adoption? Wood points out that the media focuses on Eyeborg and Mann, early solo acts who built their own devices. But there's a difference between two novelty cameras and a society of them. What changes when you have millions of AR users, wearing commercial products over which they have less control—in fact, what happens when the user base is no longer white, affluent, nerdly, and *dudes*? "There is a flip that happens between when something is rare and tactical and when it is commonplace," Wood says. "We don't know what's going to happen."

THERE IS AT LEAST ONE MORE EYEBORG waiting in the wings. Her name is Tanya Marie Vlach, and on a chilly winter day, as she sits drinking coffee in a San Francisco café, she reaches into her bag and plunks her extra eyes onto the table. There's one with a black iris etched with concentric silver rings. There's one in an iridescent mother-of-pearl; instead of an iris, it has a concavity into which she presses Silly Putty and then flowers. There's one that matches her remaining natural eye, a pale blue. She would really like a camera eye. But building the second has been harder than building the first.

Vlach lost her left eye in 2005 after a car accident. Before then, she'd been a theater manager, taking classes on the intersection between arts and computing, reading William Gibson and Philip K. Dick, which left her feeling that anything is possible. And so, she says, "as I was coming to in the hospital, morphine induced, I was like, 'Well, OK, bionic eye! That's what's next!'"

Much like Spence, her plans for a camera eye are commingled with an alternate self-identity. After her accident, she started a blog, a way of "coming up with a new version of myself, or different ways to express myself," she says, and later took screenwriting classes, which she used to draw out the details of her alternate, Isa, a cyborg assassin. Isa's bionic eye can identify faces, sense temperature, identify materials, and become a hidden stand-alone camera. But it's also networked to a command center, and that causes friction with her autonomous self. (Vlach is no slouch about acknowledging the networking issues inherent to cyborg body parts.) Thanks to her theater background, Vlach's dreams for her eye camera have a distinctly performative flavor: She imagines walking through strange cities, using the images she captures with her eye to tell her audience where to find her. She imagines roaming around in a high-tech getup and filming people's reactions. There's a concept she dubs the "gonzo futurist," using her eyecam to document protests or other live events. "The interesting thing about recording from a robotic eye is memory," she says, "and being in the middle of it, showing what you're experiencing in life, sharing."

But she sometimes envisions playing to an audience of one. Take the idea of total recall—you could explore the juxtaposition between the recording and your personal memory, working with "everything from an argument to car accidents to where you left your keys," she says. Or what about reliving life's pleasures? "I fell madly in love with this boy when I was in Madrid," she says, wistfully recalling a certain moment. "We were having hot chocolate and churros. If I could have had that documented, I would put that on a loop somewhere. It was the most beautiful sensation."

In 2008, *Wired* cofounder Kevin Kelly helped Vlach put out a call for engineers, and that brought on the same fusillade of attention Spence received, although for a woman, it had a certain icky overtone—reporters who leeringly asked whether she would film in the bathroom, and so on. "I mean, there was a lot of comparing me with Seven of Nine," the sexy *Star Trek* character who wears an implant over her eyebrow, Vlach says wryly. She launched a webpage, raised $19,000, and soon crossed paths with Grammatis and the Eyeborg team, who already had a design.

Then things bogged down. The Eyeborg team thought they could easily re-create Spence's device, which Grammatis estimates is worth about $200,000 in donated parts and labor, and took two years to develop. But Vlach's eye is different; she has a half-shell prosthetic that sits atop part of her original eye. There's less space for components. "It's like trying to cram a camcorder into a space that can fit three dimes on top of each other," Grammatis says. After laser scans and a fit study, they realized they'd need a different design, probably one that encased the parts in molded plastic. That requires more money, plus expertise, like mold making, that none of them have. Vlach doesn't have donors for parts, as Spence did. The engineering team, now older and with more responsibilities, has less free time to give.

So Vlach tried to fundraise through traditional means, like venture capital firms, but it's a hard sell, she says, because "you can't really contest that enough one-eyed people are going to purchase this." The medical industry is more interested in retinal implants—and with eyecam construction in a holding pattern until more money turns up, Vlach and Grammatis have started to think about medical applications, too. Grammatis says he has been approached by people in the defense industry interested in exploring how to couple realistic-looking eye cameras with vision-restoring prosthetics—if someone could pull that off, he says, "that's what the future looks like." For Vlach, a new dream is to work with a company or foundation that would pair engineers with disabled people to tailor-make prosthetics. "Because everybody's a little bit different," she says. "Maybe they need different things to be in the world, or to feel good about themselves." It's an idea that's right at home in the world of augmented reality, and the DIY and hacker community that surrounds it. Make it your own. Bend the rules of reality. As Mirza puts it: Self-evolve.

And so Vlach is still waiting for a lucky break, for the day when her eye, too, comes home from the shop. She doesn't *need* a camera eye, she emphasizes; she doesn't really even need a glass one. It's just hard to resist the urge to evolve. "For me, it's transformative. It's the ability of rebuilding my eye, which was lost," Vlach says. "As far as my creative impulse, I am kind of at the mercy of it. Like, *I just have to do it.*"

ELEVEN

....

New Senses

FORREST IS WAITING FOR HIS APPOINTMENT with Samppa Von Cyborg in a piercing parlor, surrounded by people whose arms and midriffs swirl with tattoos and whose faces glitter with jewelry. But Forrest (no last name; he'd like to keep it private) is conspicuously unmarked, a t-shirt-and-jeans kind of guy, with slightly mussed brown hair and an open grin. He thinks of himself as a "stealth cyborg" whose modifications aren't visible from the outside.

"In my left hand I have seven magnets," he says, flexing it open. There is one each in the tips of his pinkie, index, and middle fingers; the others are in the webbing between digits. He offers a finger to squeeze. The implant site looks like a small blister; the magnet moves like a grain of rice below the surface. With it, Forrest says, he can feel a hidden world.

"You have five senses," he says. "I have six."

Forrest hopes to expand his perception by altering his body, rather than using virtual surroundings or removable accessories. Among the biohackers who are experimenting with sensory mods, by far the most attempted new sense is magnetoreception, or perceiving magnetic fields. This ability is well distributed in nature—many species of migratory birds, sea turtles, insects, even bacteria have it, although how it works is poorly understood. Scientists haven't precluded the possibility that humans might have it, too, although it seems a very long shot. Magnet implantation is an improvisation, a hack to see if we can do an end-run around evolution. Through the magnet, could one collect—and then write in to the brain—a class of environmental information we can't otherwise perceive?

Most of the people exploring magnetic implants are not laboratory scientists; they are generally biohackers, transhumanists, and body modification artists, who work free of the restrictions ruling universities and government agencies because their exploration is considered body art, not medical research. So there isn't much official documentation of their work, only a pool of first-person accounts that's been growing since the turn of the century, when the implants began taking off. These magnets aren't particularly strong—not powerful enough to wipe credit cards or computers. Most people get one or two, not a handful. But people like Forrest say they can now feel what they otherwise could not.

"Bad analogy, but in *The Matrix* it's kind of like taking the pill," Forrest says, referring to the scene in which the hero must choose whether he wants to stick with his ordinary life or see the world as it really is. "You wake up and you're using your hand and it's a whole new world there and you've got to deal with it. Sometimes it's nice. Sometimes it's not. It does get annoying at times. But it's there, and there's no avoiding it. It's your new reality."

With his magnets, he says, he can feel another layer of information in his environment. There's hidden metal: the screws in a deck, a staple in paper, wire in the wall. He can feel, of course, the tug of other nearby magnets. But the main invisible force that he and most others report sensing is electric current: in outlets, in fluorescent lights, in power cords. He describes the sensation as like an electric shock, but "the buzz without the bite." His weirdest experience so far, he says, was walking through a security door at a store and feeling his hand be drawn toward its alarm system. But most of the sensations are pleasurable, like the vibrations from movie theater speakers. "It's amazing," he says. "You are actually feeling the sound. The big rumble, it runs through you."

That said, it's not enough. So he's back for more.

Forrest is called into a small room with a reclining chair, like a dentist's. A sterile blue paper drape lies over its seat. Forrest takes a seat next to it and stretches his arm across the drape. Samppa Von Cyborg sits across from him and attentively listens as Forrest describes where he'd liked the magnets placed.

Von Cyborg is a commanding, almost patrician, presence, a veteran in the body modification world. He has been piercing since the 1990s. Originally from Finland, he is now based in London, and travels the world doing modifications by appointment. His head is shaved and on this day was covered by a Harley Davidson bandana, but in the past he's sported a "metal Mohawk," a double row of metal spikes down his skull. His teeth are covered by etched metal caps, his ears stretched and shot through with rings, his body a tapestry of tattoo work. Both he and his partner Aneta Von Cyborg, who assists with his procedures, have a series of "ribs" inset in their right arms, subdermal silicone implants that coil from wrist to elbow, as though they were wearing beaded bracelets below the skin. Among Von Cyborg's contributions to the trade are the Mad Max Bar, a piercing that zig-zags in and out through the skin, and a technique for shaping the top of the ear into an elf-like point.

Von Cyborg started experimenting with magnets in the early years of the 2000s, including trying a few himself, but wasn't too impressed—the first magnets he tried were spheres, so they would just spin in the scar tissue pocket that formed around them, and didn't produce much sensation. He restarted in earnest in 2011, after rolling out a new design he termed "super strong" magnets, the kind Forrest will get today. Right now they're soaking in a sterilizing fluid; they're pellet shaped, iron gray, 3 by 7 millimeters each, coated to protect them from disintegrating in the body. As Forrest explains where he wants them to go, Von Cyborg uses a felt-tip marker to draw on his skin: three in the fingertips, one in the base of his right palm, one near the crook of his elbow.

The piercing itself is swift and nearly bloodless; at each spot, Von Cyborg makes a small opening, pushes the magnet through, presses the skin firmly with a cotton swab, bandages it. The implants aren't deep, Von Cyborg says; they are "in the subcutaneous layer, so just under the skin." (Kids, don't try this at home. This is fast because it's done by professionals with special magnets, not those you can find at the store.) Forrest grimaces mightily throughout, his eyes closed as he tries to keep his arm perfectly still. When it's over, he's delighted. "It was intense," he says, as everyone heads outside for a smoke.

Von Cyborg cautions that it will be a couple of months before Forrest knows how sensitive his new magnets will be; he'll have to wait for his hands to heal. But other than owning a dozen magnets instead of a couple, he says, Forrest is typical of the magnet implant clientele. While the early adopters were long-time body modders interested in party tricks—like dangling paper clips or bottle caps off their hands—or in larger versions that could be seen through the skin, today most people are in it for exploration, not self-expression. "It's not visible; it's completely just for yourself," Von Cyborg says, as he takes a thoughtful drag on his cigarette. "It's a lot of geeks who want this kind of implant, because it gives you this extra sense."

ANYONE WITH A FUNCTIONING SENSE of skepticism is going to wonder whether putting a magnet in your hand can produce true magnetoreception—or, in fact, if any hack can extend sensation beyond the five known human modalities.

Here is the biohackers' argument in favor of trying: Magnetoreception exists in nature, so evolution has already built this biological machinery. We know that other animals can write this signal from their environment into the brain, which converts it into meaningful information. Even if we aren't born with the right sensory gear, why can't we mimic this process with a clever workaround? The brain is remarkably plastic; the history of retinal and co-chlear implants among the formerly blind and deaf shows us that people can learn to separate signal from noise, even when adapting to a class of information that is entirely new to them. And even if there are no studies, the feedback from magnet implant recipients has been remarkably consistent: People say they feel electrical currents, the pull of other magnets, metal objects. It's hard to write them all off as simply wishful thinkers.

That said, we are not born with magnets in our hands, or any obvious magnetoreceptor, nor is there a sensory area in the brain devoted to magnetoreception, as there is for hearing and sight. There is no behavioral proof that humans have a functional magnetic sense in ordinary life. It's also not clear how information collected by an inserted magnet would be conveyed to the brain, and how the brain might process it. If we have no magnetoreceptors, then the magnet's effect must be indirect, acting upon receptors and nerves built for other modalities. In that case, is it really magnetoreception, or something else?

By the end of this chapter, I'll raise a hypothesis about what is behind the perceptual experience people like Forrest describe. But first, let's look at the really big and extremely cool problem this biohacking cottage industry is trying to confront: Our collective angst about missing out, our desire to know if our sensory world could be *bigger and more awesome.*

Because it is, in fact, frustratingly small. We can't perceive everything, because our sensory organs work within limited parameters. Our eyes only see part of the electromagnetic spectrum, the visible wavelengths between about 400 and 700 nanometers, but not ultraviolet, infrared, gamma, or X-rays. Our ears only hear frequencies in the 20 to 20,000 Hz acoustic range. Yet some animals don't share these limits. Honeybees see UV light, which helps them find markings on flowers. Bats, dogs, dolphins, and some insects hear in the ultrasound range. These are not separate modalities, or new senses. They are larger ranges for senses people already have: vision and hearing.

Some animals use sensory information in ways humans generally cannot. For example, bats, whales, dolphins, shrews, and some birds use echolocation or biological sonar to sense distance in low-visibility environments like caves, underwater or underground. They can navigate or look for food by using the sounds bouncing off objects as spatial reference points. (Echolocation is using one's own vocalizations to produce sonic feedback; sonar is using other noises.) This is a specialized use of hearing, not a separate modality. While the ability isn't inborn for people, there are a few well-documented instances of people learning to do it, notably Daniel Kish, the founder of the nonprofit World Access for the Blind, who teaches a method he calls "FlashSonar." By clicking his tongue, he can ascertain nearby objects' location, their size, and sometimes even their material.

And finally, some animals have sensory ranges and even entire modalities we don't, because they have body parts we don't. Sharks and rays can sense electricity, and use it to detect objects and other animals in low-light conditions. Their bodies contain internal canals filled with a conductive gel, which lead to special cells that detect changes in voltage. Some snakes sense infrared radiation, which follows the visible red light wavelengths on the electromagnetic spectrum. They transduce this information through a facial "pit organ," part of their heat-sensing touch or trigeminal system. (In people, the trigeminal nerve near the mouth conveys information about chemical irritation and mouthfeel.)

People probably don't have the right parts for magnetoreception, either. But it's hard to say for sure, because we don't know what gear animals are using. "The biggest unanswered question in sensory biology is, how does magnetoreception work?" says Duke University sensory ecologist Dr. Sonke Johnsen, who studies vision and magnetic sensing in sea animals.

It's a tough proposition for some really confounding reasons. First, Johnsen says, with the other senses, even if the receptors are small—like the photoreceptors in the retina or the hair cells in the cochlea—they are surrounded by a larger organ that collects and amplifies signals. "None of this exists for magnetoreception," Johnsen says. Because magnetic fields don't interact with biological tissue, as light and sound do, nature didn't build a focusing organ. Without one, it's hard to find the right circuit to study. "You are just looking for a handful of neural cells," Johnsen says. Not only could they be completely nondescript, but they could be anywhere—they don't have to be on the body's surface, as other sensory organs are, or even in the brain.

Wherever they are, these cells must convert an electromagnetic stimulus into an electrical signal that can be read by the brain. Then, Johnsen says, animals likely use this information in at least one of two ways: as a "compass" that tells which direction they are headed, or as a "map" that tells them where they are geographically, likely through sensing the degree of inclination of the Earth's magnetic field lines. These lines form giant arcs between the planet's poles, creating angles the animal can sense. It's unclear what this perceptual experience might be like to them—after all, try asking a sea turtle. Studies have suggested some birds see inclination lines superimposed over their vision, but for other species it could be an entirely unique sensation.

Right now, there are three theories of how magnetoreception works. All might be correct, Johnsen says—in fact, animals might use several mechanisms to get around. The first hypothesis is electromagnetic induction, and it's specific to sharks and rays. Because their bodies conduct electricity, as they swim through magnetic fields, their electrosensors may be able to detect voltage changes that vary depending on their orientation to the fields. But it's hard to prove, says Johnsen, because it's hard to design a behavioral test that separates these two abilities. (Also, it is just hard to design behavioral tests for sharks.)

The second, the magnetite hypothesis, is known to work in magnetotactic bacteria, which use this ability to orient themselves to a certain depth in their muddy habitat. They form crystals of the mineral magnetite inside their cells, which—like "a tiny little compass needle in your cell," says Johnsen—rotate to line up with the Earth's magnetic field. This movement might trigger a neural reaction either by opening an ion channel in the cell or by putting pressure on some other receptor mechanism. A similar process has been studied in fish and birds. But just having magnetite doesn't mean an animal is magnetoreceptive—these particles turn up everywhere, including in the brains of older humans. "Magnetite is iron oxide, and it's a common breakdown product of hemoglobin," Johnsen says. "There's magnetite all over the place. It doesn't mean that it's part of any receptor system. It just means it's there."

A third possibility is the radical pair hypothesis, and it's trickier to unpack. Dr. Steven Reppert, a neurobiologist at the University of Massachusetts Medical School, has made a case for it in fruit flies and butterflies. This theory suggests a chemically based system: Light-sensitive molecules set off an intramolecular chain reaction that generates radical pairs, or unpaired electrons with alternating spin states. This alternation creates a magnetic moment, Reppert says, "and that could sense the very small changes in the Earth's magnetic field." The idea was proposed by physicist Klaus Schulten at the University of Illinois at Urbana-Champaign. Working with Schulten, Thorsten Ritz of UC Irvine suggested the light-sensitive molecule is a photoreceptor protein called a cryptochrome, usually found in plants and animals as part of the circadian clock system.

While much magnetoreception research has focused on the behavior of birds and sea turtles, Reppert studies cryptochromes in species that are easier to genetically manipulate. In a series of experiments, he showed that fruit flies in a maze naturally avoided a magnet positioned next to it, but could be trained to fly toward it if given a sugar reward, indicating the flies sensed the magnet. Crucially, this only worked under ultraviolet-blue light, specifically the 380- to 420-nanometer wavelengths—"and cryptochrome is felt to be a blue light photoreceptor, so this made sense," says Reppert. But it didn't work in flies without the cryptochrome gene, says Reppert, "so that proved cryptochrome was involved."

Then they experimented with monarch butterflies, a migratory species. Reppert says monarchs probably use several navigational mechanisms as they fly south for the winter: On sunny days they use the horizontal direction of the sun, and "on partly cloudy days they can use the polarized sky light pattern by seeing a patch of blue sky." But the mystery is how they keep going when it's overcast, he says, "and the magnetic compass fit that bill."

Reppert's lab uses a flight simulator for butterflies, a large barrel in which the flight orientation of tethered monarchs can be monitored. Surrounding the barrel is a magnetic coil that can mimic the inclination angles of the Earth's magnetic field. Butterflies exposed to an inclination mimicking what they'd encounter while flying south do indeed orient south. But when the lab reversed the field's inclination, the butterflies oriented north. "That is the acid test for an inclination compass," says Reppert, showing butterflies were magnetically sensitive. Then, using light filters, the lab found it only worked in the same ultraviolet-blue light wavelengths they'd seen with the fruit flies, suggesting cryptochromes were involved here, too.

In monarchs, Reppert believes the cryptochromes used for magnetic sensing are primarily in the antennae. So next his lab painted some of the butterflies' antennae black, to prevent light from getting to the receptors. The butterflies with black paint flew in confused circles, while another group with clear paint oriented normally. So taken all together: Two insect species show magnetoreception under the blue light that cryptochromes sense, and these responses are blocked when the receptors are obscured or the related genes are deleted. But exactly how radical pairs might be generated from this process is still debated, Reppert says: "It's all hypothetical."

The most passionate case for human magnetoreception was made in the 1980s by University of Manchester zoologist Robin Baker. In one of his studies, blindfolded students were driven far from campus via a "torturous route," then (still blindfolded) asked to guess which direction they were in relative to their school. In a subset of the tests, some students wore a bar magnet inside their blindfolds, while others wore a placebo (a brass bar). Baker concluded that "humans have an ability to recognize homeward direction even in the absence of visual cues," and those wearing a magnet performed worse than the placebo group, which "suggests some influence of the bar magnets on orientation." He also studied how well blindfolded students, wearing magnets or brass bars, could guess their direction after being spun in a chair. He argued for a version of the magnetite theory, suggesting that humans might have deposits of magnetic material in their sinuses. But no one has proven it, or repeated his results.

Both Johnsen and Reppert are dubious about the prospect of human magnetoreception, but they won't rule it out. "There is no convincing behavioral evidence that we can do it," says Johnsen. That said, he continues, "A reasonable number of animals can do it, and so it's not impossible to suggest that humans can still do it. But we don't seem to be able to access it. It could be very vestigial."

In 2011, Reppert's lab published a paper regarding cryptochromes found in the human retina, which are considered related to circadian clocks, not magnetoreception. Yet Reppert's study showed that the human gene, when inserted into fruit flies bred without their own gene, allowed the flies to sense magnetic fields. That shows the human form "has the molecular capability of acting in a magnetosensing system," Reppert says. "Whether it actually does in the eye of humans, we don't know. But it suggests that maybe it's something we should look at again."

Both scientists were amazed to learn that people are inserting magnets into their hands, and both had the same reaction: We can't know if it works until someone does a controlled experiment. Johnsen noted its similarity to the magnetite theory, in which a particle moves inside a cell, although on a larger scale. That said, while he has no doubt people can feel an embedded magnet move in response to a large nearby magnet, he says it would be impossible to use one to read the Earth's magnetic fields, as birds or butterflies do. "The Earth's magnetic field is so weak. It just does not have enough oomph to create any real turning torque," he says.

"I think that's a pretty wild idea," Reppert says with a chuckle when he hears about the implants, but then again, thanks to that human cryptochrome paper, he's still getting emails from people who claim to have a magnetic sense. "So nothing surprises me, and this clearly doesn't," he says good-naturedly. "Will it work? Well, good luck. I guess the experiment is already going on."

A GOOD DEAL OF THAT EXPERIMENT is going on at Grindhouse, where getting a magnet is just the first step in sensory exploration. (They jokingly call it "the blood sacrifice to the grinder gods.") On a late weekend night, the members have gathered in the living room, perched around a coffee table on which they've laid out the floor model of the Circadia, an implant they built to read out the body's internal metrics.

Its twin is in Tim Cannon's arm. Cannon, a software developer by day, is one of the group's founders, and its most vocal member: His delivery style is brash, speedy, hyperintelligent. Despite the Pennsylvania winter, he is barefoot and outfitted in cargo shorts and a t-shirt that declares, "This Is What

Awesome Looks Like." He's had the Circadia for about three weeks now, a lump slightly smaller than a deck of cards on the inner plane of his left arm, just below the elbow. The surrounding flesh is swollen and pinkish, but otherwise seems healthy; he can touch it without flinching. The whole thing is submerged beneath a tattoo of a DNA double helix encased in a gear.

Shawn Sarver, more dry and reserved, is the group's electrical designer, a former Air Force avionics technician who now works as a barber and drops by most days after work to build stuff in the basement. He gestures toward Circadia's demo version, which like its twin is encased in a clear silicone shell, showcasing the parts inside: a receiving coil for wireless charging, a battery, a Bluetooth module, a row of LEDs. "This was more proof of concept, but you can get these LEDs to show through your skin and you can communicate data this way," Sarver says. To demonstrate, Cannon picks up a power-charging coil and holds it to his arm. From beneath his skin, the LEDs flash green three times, then sustain a steady red to show the device is charging. Someone starts humming "Jingle Bells" as we watch the Christmas-y glow of Cannon's arm.

"Well," he says wryly, "I'm a festive guy."

There's another reason they made the device light up, and that's to prompt public reaction: Maybe the people who see it flashing will want one, or dream up new artistic possibilities for it, or simply shake their heads wondering what this dope did to his arm. The point is to be the opposite of a stealth cyborg—to be very obviously modified. "It really illustrates that the cyborg future is here," Cannon says. "We really want people to understand, holy shit, that guy's got a piece of electronics that he built under his skin, taking data and it's Bluetooth enabled and so he's controlling it with his phone. We have *merged*!"

Grindhouse—Cannon's house, really—is a small brick two-story on the edge of a Pittsburgh suburb. Sarver likes to point out that all the roadkill in town seems to stack up in front of it; it's the line where civilization meets wilderness. And that seems about right. It's a workspace for a collective of biohackers—they were up to 17 at that point—although only a few were living in the space that winter, including Lucas Dimoveo, who was curled up in an armchair, pecking at a laptop as the others showed off the Circadia. Dimoveo was running operations for the group after dropping out from his biology degree program, more inspired by possibilities than the already extant. The group's name derives from the term "grinder," or body hacker, itself borrowed from Warren Ellis' *Doktor Sleepless* graphic novels, which portray a future in which "this marginalized impoverished group of wild punk rocker people basically take it upon themselves to build their own augments," says

Cannon. Frustrated with the slowness of progress, they spray paint a slogan that is now the Grindhouse mantra: *"Where's my fucking jet pack?"*

The group formed as Grindhouse Wetware in early 2012 to develop and eventually sell their own devices. So far they've invented a handful of wearable gadgets that work in conjunction with implanted magnets. The Circadia is their first implant. There's only one sensor in this version, for body temperature, although they've imagined future iterations that could measure heart rate, blood pressure, maybe even blood oxygen, porting data to your phone so you can check your vitals. "I would love to be able to go, 'Hmm, oh yeah, the heart's doing just fine,'" Sarver says, in the casual tone of someone checking his watch. Or maybe objects in your environment could read your data, so your thermostat reacts to your temperature, or your house dims the lights if your blood pressure's up after a bad day. All of that is pretty far off from what this Circadia can do, which at the moment is mostly light up; they can't try reading Cannon's temperature until the swelling goes down.

Cannon and Sarver became friends thanks to their complementary skill sets; Cannon knows software and Sarver knows hardware. They'd tried a few hobby projects, says Sarver, and then "Tim one day found this article on the Internet where this woman by the name of Lepht Anonym had implanted a magnet in her finger and it allowed her to feel electromagnetic fields." Anonym, a Scottish college student and blogger interested in the prospect of sensory extension, wrote extensively about her experiments as a "scrapheap transhumanist," implanting herself with radio-frequency identification (RFID) chips, magnets, and an attempt at a temperature sensor. Even though she openly shared some fairly horrific details—hospital visits, infections, and the pain and mess of digging items both into and out of her skin, sometimes using a vegetable peeler—she got it done. And the magnets worked. "And that got Tim really excited," recalls Sarver. "And then a month later he has one. And then he's like, 'Let's build things to interact with this.'"

Dimoveo was similarly inspired by Anonym. He is one of the founders of the biohack.me web forum, where people meet to discuss "functional (sometimes extreme) body modification." But by the time he was a college freshman, he had become disenchanted with how slowly development was going, mainly because people could not figure out how to power implanted devices. He created a post calling for people ready to get to work; Cannon responded, then looped in Sarver. At one of their first meetings Dimoveo proposed creating hardware that would interact with an implanted magnet. "I remember Shawn and Tim looking at each other and going, 'We finished that three months ago and it's in our closet!'" Dimoveo recalls. "So that's where it started."

That first invention is the Bottlenose, and it links with an embedded finger magnet to create a kind of sonar. It's a small plastic box containing an ultrasonic range finder attached to an exterior coil, magnetic wire wrapped around a steel nail. In a demo filmed in the Grindhouse basement, a blindfolded Cannon holds the Bottlenose in his hand, and uses the vibrations transmitted from coil to magnet to find objects (mostly cereal boxes) and locate where Sarver is standing, and to tell whether they are moving toward or away from him. They also worked on the Thinking Cap, a hat with built-in electrodes to deliver current to the brain, and an abandoned project they called the Sixth Sense Glove that would let a computer track your hand motion.

From there, they wanted to jump immediately to an implantable compass. This idea had its genesis in the North Paw, developed by San Francisco's Sensebridge hacker group, which is worn around the ankle and vibrates motors to let you know which way is north. Anonym planned to make a subdermal version dubbed the Southpaw for her left leg, but hasn't finished. The Grindhouse version, which they've called the Northstar, is meant to go in the back of the hand. Their initial plan was to use an embedded magnetometer to sense direction and for the star-shaped implant to light up more brightly when you point north, providing a semi-intuitive sense of direction. But they quickly realized they needed to start with something simpler. Hence Circadia. "This is our learning tool," says Sarver, and it is designed to answer a few crucial questions: How do you put electronics in a body and wirelessly recharge them? How can you port biometric data to a phone? How can you do it all without killing Tim?

At the moment, they're not entirely sure if they're killing Tim. The Circadia's not exactly comfortable; they'll have to make future devices smaller and rounder. And although they extensively tested the parts, including trying to hammer a nail through the battery to see if it would crack, Cannon is constantly imagining the worst. "Every time I have a little headache, it's because there is some sort of lithium polymer seeping into my brain," he says. "Every muscle twitch, the battery has breached, it's shorted, it's minutes until I'm going to die!" They planned to leave Circadia in for six months, then take it out to assess wear and tear. (Ultimately, they took it out after three, after they realized the battery was expanding from the heat from the charging coil. The silicone stayed intact, so Cannon suffered no ill effects.) But this, they feel, is the risk you have to take if you want a product nobody else is making.

Cannon's implant was performed at a festival in Germany by body modification artist Steve Haworth, although Cannon couldn't bear to watch. ("I looked once and I regretted it," he says.) Their alliance with the body art community is a practical one; a piercer can easily do simple jewelry implants that would raise confounding insurance and medical necessity questions for a doctor. But Grindhouse folks approach modification more as a vehicle for exploration than self-expression; in fact, says Dimoveo, "one of the big things that brought us together was perception. We were all interested in creating or adding different senses." It's the "Where's my fucking jet pack?" ethos applied to biological systems, a feeling that humans are unfairly missing out, that the world is open to the pit viper and the shark and the butterfly in a way that is closed to us. "Humans are very ambitious. We want to explore everything," continues Dimoveo. "There's so much of the universe right here and now that's just passing by us." Cannon says getting his magnet, and feeling something he could not before, was a sort of wake-up call: "It's a profound sense of how blind you really are."

That said, everyone in the group has their own angle on which sensory portal they'd like to open. "My dream augmentation is getting ultraviolet and infrared photoreceptors," says Dimoveo. "This sounds kind of sappy, but I'm a big fan of sunrises, so I always wonder what it would look like in ultraviolet and infrared."

"I want to feel the entire electromagnetic spectrum," says Sarver. "I want to feel something when there's X-rays going on. I want to feel if there's any gamma ray bursts somewhere. I am interested in knowing all of that, because that's all reality, and I am only getting a tiny little slice of reality."

"For me it's a little different," says Cannon. "I feel like I'm trapped in a biological body with a shelf life." He's in his early 30s, so this is not a midlife crisis, he says, more like a desire for his lifespan to be decoupled from his bacon intake, or a perpetual existential dread that he worries everyone else might be ignoring. "We're all going to die eventually!" he says in frustration. "Doesn't that *alarm anybody*, that we're sitting in a vessel that's on its way to death and sweet oblivion?"

All of them, in fact, are pretty disenchanted with the body. Dimoveo has a particular reason to feel disaffected; then 21, he was already completely bald due to alopecia, an autoimmune disease that causes hair loss, sometimes including the protective cilia in the nose. "I've always had poor health and got really strange weird diseases," he says, including surviving a serious lung infection. "Growing up, I very quickly became aware of how crappy the human body can be."

"For me, it is not as much about augmenting the body so much as getting rid of it altogether," Cannon says. "I'm extremely underwhelmed in terms of—"

"Capability," Sarver puts in.

"If you look at like the brain, it's obviously not designed," says Cannon.

"No one competent would engineer it," adds Dimoveo.

"When you look at biology, it's clear that nobody's at the wheel," Cannon concludes. "For me, it seems like it's a bad system for supporting consciousness."

That said, they have great faith in humanity's ability to collectively engineer its way out of biological problems. Maybe Grindhouse can't build a better brain, or a bacon-resistant heart, or an immortal body, at least not with the parts they just brought back from one of their nighttime pilgrimages to RadioShack. But perhaps they can cobble together a sensory device that will let them peer into the universe of information that is passing them by. So they uncurl themselves from around the coffee table and head down into the basement to see what they can make next.

MUCH OF TODAY'S SENSORY EXPERIMENTATION begins with the magnet, and the magnet begins with Steve Haworth.

Haworth is a soft-spoken man with a shaved head and three ghostly slashes cutting across his right eye, like the scars of someone who's had a run-in with a puma. "When they were first done they were deep and bloody and gash-y, but they just healed beautifully, unfortunately," he says ruefully, wishing this bit of body art had remained more clearly visible. When he's not traveling doing modifications or teaching, he works out of his Arizona home, where he's built a piercing studio and a clean room for manufacturing body jewelry. Haworth's background is in medical device and implant design; he designed cannulas for surgery before moving into body modification in the 1990s. His ·inventions include ear punches, which cookie cutter a shape out of the upper ear cartilage, silicone implants—often shaped like hearts or brass knuckles or hand grenades—that rest just below the skin, and techniques for transdermal implants, in which a post anchored below the skin protrudes through it so that jewelry can be attached.

Haworth's been exploring magnetic implants since around 1999, when he and body jewelry designer Jesse Jarrell came up with the idea of using implanted magnets to anchor a watch, or perhaps spikes, or a goggle shield over the eyes. As a first run, they implanted magnets on Jarrell's wrist, and then six weeks later clicked on a watch. It stuck, but there was a problem. The magnets were pressing together too tightly. "If you kept it on for more than 20 minutes," Haworth says, "you could potentially cause your tissue to go necrotic through starving it of blood."

So that idea was out, but they came up with a new one after Haworth met a man who had accidentally gotten a piece of tool steel stuck in his pinkie. "He was working in a speaker manufacturing plant, and he was actually able to go by and tell which speakers had been magnetized and which ones hadn't," says Haworth. If they reversed the process and put a small magnet in the finger, they thought, you'd be able to sense other magnets.

His first volunteer was Todd Huffman, who was studying bioinformatics at Arizona State University after doing his undergrad work in neuroscience. Huffman had previously researched cochlear implants, and was fascinated by how wearers adapt to new sensory information. He'd started to wonder whether the brain might adapt differently to external artifacts versus internal ones—the difference between, say, wearing headphones and having a cochlear implant. If a sensory device is temporary and removable, he figured, your brain doesn't "internalize the mental model the same way that you do a sense that is part of your body." So when it came to magnets, Huffman thought, maybe you could try gluing one to your fingernail, but it wouldn't be the same as putting one inside your body. "What I wanted to do was go through the first-person experience of trying to integrate a new sensory modality into my worldview," he says.

The three of them came up with a magnet design, expecting Huffman to mostly be able to feel other magnets. But soon after his implant, while reaching for a frying pan, Huffman was shocked to feel something else: the magnetic field thrown off by his electric stovetop. That moment of registering a new force was what he'd been looking for, he says, "breaking my mental model of reality with a new sensory experience."

Haworth manufactures his own finger magnet jewelry, different in design and implantation technique than Samppa Von Cyborg's. His are round, gold-colored pucks about the size of an aspirin, surrounded by a silicone coating. He is extremely open about the safety concerns surrounding implanted magnets, pointing out that his first six test implants failed and had to be removed. These were dip coated, which created a thin spot in the casing and ruptured, allowing the magnets to disintegrate. So Haworth came up with a new manufacturing method involving injection molding and curing at 2,000 psi. "Under that pressure," he says, "the crystalline structure locks in incredibly tightly. It's almost impossible without a sharp object to break one of these encasements." And the trend took off. Haworth estimates that by early 2014 he'd done about 3,000 magnet implants, and sold another 5,000 magnets. Among these, he knows of only four failures.

Haworth and Huffman agree the sensory experience you get with a magnet isn't very complex, and although some people started using the term "magnetic vision" to describe the experience, it's not actually visual. "As a sixth sense, it's not as unique as sight or hearing," Haworth says. "But just like the hairs on your arm make you aware that something moved by your arm, magnetic vision makes you aware that you've passed through a magnetic field," or the area surrounding a magnet or electric current.

Haworth says it's hard to explain what it feels like if you don't have an implant; he tried taping magnets to his nonimplanted hand and comparing sensations because he wanted to know if a magnetic ring would produce an identical effect. But, he says, it's just not the same. Still, he has some workarounds. "Let me see the tip of your finger," he says, pressing mine against his own middle finger, where his magnet is. Then he picks up a large magnetic ball, and brings it close to our hands. From below his skin, I can feel his implant flutter. Haworth says because it's puck shaped and in a smooth encasement, it's flipping end-over-end within its scar tissue pocket.

Then we walk into the kitchen, and once again press fingers as he turns on the electric can opener. This time, I can feel the magnet hum beneath his skin. "That's it vibrating to the oscillating magnetic field," he says. He moves his hand away from the machine, then close again, tracing the cardioid trajectory of the field. As he and his partner Mandi Vaterlaus, who joins us to demonstrate her own, much stronger magnet, curve their hands through the air, I ask if they think they're feeling a unique modality, or simply touch. After all, the magnets are interacting with nerves in the skin, because their movement creates mechanical pressure. I'd also been struck by Haworth's analogy of something moving the hairs on your arm—a tactile metaphor.

This is a tough question; by the time I got to Haworth's house, I'd asked it of a dozen magnet implantees. Nearly everyone had struggled to articulate a feeling that seemed to be both touch and not, sometimes describing it as synesthesia, or crossover between two senses. "It's similar to touch. It just reacts differently," Haworth says thoughtfully.

"It's similar in that you can kind of get a concept of where [the field] is from start to end, and where it's stronger, and you can kind of get this 3-D map in your mind of how it works," says Vaterlaus. "But it's not quite the same as touch because you can't really reach out and grab it. You can't manipulate it with your fingers."

"It's like a heat source," Haworth agrees. "The closer you get to the heat source, the more intense the heat. The closer you get to the magnetic source, the more intense the magnetism."

Like most people with an implanted magnet, they felt they had gradually adapted to it. Haworth attributes this partly to the thinning of the scar tissue, but also to the brain learning to recognize novel input. "After six months you get very used to it," he says. "And it just becomes part of you. It doesn't become alien or strange. It just becomes part of your subconscious."

People with magnet implants often tell anecdotes about subconscious reactions: involuntarily wincing when seeing others do things that would be painful to someone with a magnet, dreaming of magnetic sensations, reacting in genuine surprise when they unexpectedly come across something magnetic. For example, Haworth says, one day he was cleaning up a pile of dolls belonging to his then-girlfriend's daughter, "and all of the sudden I pick up this one Barbie doll and my finger is getting pulled towards it, and I'm like, 'Wait! *What?*'" The doll turned out to have a magnet inset in her crotch to attach a plastic bathing suit—definitely not a place where anyone would have primed themselves to expect a magnetic reaction.

Most of all, Haworth says, to him it's a new sense because he reliably uses it to interface with the world in a way he otherwise cannot: it's something he notices every time his hard drive spools up, when he runs the microwave, when he's in his shop working on electronics. "I've been doing this for a long time now, and if I lost my magnet I would literally feel like I lost one of my senses," Haworth says. "There would be something significant missing from my life."

ASKING A GROUP OF BIOHACKERS who among them has a magnet is a bit like asking a bunch of tech geeks who owns a smartphone: *Everyone*. They're a sort of gateway drug for a psychonaut tribe that is trying very diligently to move on to harder stuff.

In an apartment above a sushi joint in San Francisco, a dozen biohackers have gathered after a transhumanism conference. One of them is Rich Lee, who had Haworth implant magnets in a part of his ears called the tragus, the flap of cartilage that sticks out in front of the auditory canal. He wears a coil around his neck that plugs into his cell phone. It transmits sound to the coil, which creates a magnetic field, which causes the magnets to vibrate, creating sound he can hear. Also at the table are members of the group Science for the Masses, who have devised an experiment to test whether they can give themselves near infrared vision. They will carefully control their diet to eliminate vitamin A, and will instead take supplements of vitamin A2, which is linked to near infrared vision in freshwater fish species. After three months on this diet, plus cyclic doses of retinoic acid, they will test their success using a home-brewed device they've just dubbed the "Gibson Rift," a sort of Oculus-shaped headset that flashes near infrared light at the eye. They'll measure the eye's response by wearing a special electrode that goes on its surface.

Like Grindhouse members, the people at this table are driven by a desire to explore hidden dimensions. "Just knowing that there's an invisible world out there that you can't observe is infuriating to me," says Lee, the man with the tragus magnets. "We could be listening to the death rattle of the supernova that's happening right now. I think that would be amazing. Or listen to the music of the spheres, or submarine communication. There is all kinds of awesome stuff going that is just outside of our range of perception." And they are similarly unthrilled with the limits of the body. "Evolution doesn't mean that you're the most awesome thing on the planet," says Gabriel Licina, a member of the infrared project. "It just means that you *haven't particularly died as much as everything else.*" There's room for improvement, he says, and you might as well try new stuff, because even if only one experiment in a thousand works, "well, you have just owned evolution."

One of the ways they are trying to own evolution is happening at Amal Graafstra's end of the table, where he is about to put an RFID chip into the hand of a young woman named Erin. Graafstra runs the company Dangerous Things, which sells supplies and safety guides for RFID implants; he's had two of his own since 2005. RFID chips are not sensory devices; they enable you to act, rather than feel. But they were one of the first biohacker projects to show that you could safely live with an implant long term. Graafstra uses his to open doors, identify himself to his computer, and start his motorcycle. You might already have RFIDs in your life in the form of workplace entry cards, bridge toll passes, or tracking chips in your pets.

The implant only takes a few seconds; Graafstra tents the skin in the divot between Erin's thumb and forefinger, and then uses a preloaded assembly— imagine a hollow piercing needle attached to a syringe plunger—to push the tag just below her skin. Then he shows her how to use it. He takes out a demo device—an electronic door bolt connected to an RFID access controller. Graafstra has Erin wave her hand in front of its reader to add her tag to the system. "Take it away and put it back," he says. She waves her hand again, and this time the lock snaps open. "*Cool,*" she says happily.

Graafstra didn't invent the idea of chipping humans. In fact, cyborg pioneer Kevin Warwick's first implant back in 1998 was an RFID chip. He'd worn his for nine days, using it to interact with his university building: locks, light switches, a computer that displayed his home page when he walked past, and another that logged his entrances and exits, which, he wrote in his autobiography, was of obvious interest to reporters because of "the Big Brother issue." In 2004, VeriChip turned tagging into an FDA-approved business, although the device, marketed as a form of medical patient identification, was controversial, raising concerns about tracking and hacking, and was ultimately discontinued. Several states, worried that employers might require the chips, promptly banned mandatory ID implants.

As with AR devices, implants raise tough questions about surveillance. The more intimately you wear a tracker, the more information it can convey, not just about where you are, but what you are doing. Adam Wood of Stop the Cyborgs, for example, had brought up the intrusive possibilities posed by fitness wristbands and cell phone apps that monitor not only movement but heart rate, calories burned, and how much you've slept. Maybe you like to track these metrics for your own use, he mused, but what happens if your workplace asks you to wear one, because their insurer would like to use information about collective employee health to calculate the group rate?

Implanted chips or sensors like the Circadia are meant to be worn continuously—you can't take them out or turn them off as you would a wristband or smartwatch. Gregor Wolbring, the ability studies expert from the University of Calgary, points out that with sophisticated-enough internal sensors, you could track not only a person's behavior, but biochemistry. "The surveillance will be *amazing*, to be cynical," he says. "If we are thinking the NSA thing was a problem because they are listening to some phone calls or some emails, sensors can in essence monitor *anything*, any changes in biochemical compounds in your body." Imagine sensors that track chemicals like dopamine or serotonin, or the presence of drugs in your bloodstream—illicit ones or a prescribed regimen. Whether this seems helpful or scary to you might depend on where that information is going: Just you? Your doctor? Your employer? The police? "There will be no body privacy anymore," Wolbring says.

For Graafstra and the others around the table, these particular RFID tags don't pose a threat: Their range is about one inch, and because the user controls which devices can read them, the user also controls who collects the data. For example, says Graafstra, if you install a reader to make your chip talk to your front door, "that data doesn't go anywhere. It's not going to a credit card company that knows where you purchased something and for how much and when. It's not a telephone that's in your pocket that's constantly reporting your location. It's your own data, and you keep it and you choose to share it or not."

Information access and control are central issues within the biohacker community. As a rule, they believe everyone should have access to scientific knowledge and the tools of discovery. They see themselves as a counterweight to the concentration of resources in the hands of a few—corporations, governments, universities—who, for practical and ethical reasons, put their efforts into developing products for people with common illnesses and disabilities, not the smaller slice of otherwise healthy people who want enhancements. "That was one of my biggest aggravations with dealing with the medical community," says Lee, who came up with his tragus implant idea after losing vision in one eye and, thanks to a misdiagnosis, being told that he could lose it in the other. Enhanced hearing, he thought, could be a form of compensation. But Lee wasn't blind *yet*, and he was seeking a hearing augment, not a treatment for vision. So, he says, the response was sometimes angry: "They're like, 'What? You're not broken! We're not going to fix you.'"

"This is really the difference between modern medicine and things like transhumanism and sensory enhancement and grinding," Lee continues. "One is to restore lost function and the other is to go above and beyond."

Grinders also argue that you can do what you want with your body because it's *yours*; you could call this body sovereignty, bodily integrity or self-ownership. U.S. courts have upheld related arguments over women's access to birth control and right to privacy when seeking an abortion—both acts of reproductive autonomy. They have ruled that tattooing is a form of free speech, and generally interpreted the First Amendment to cover dress and hair color, if not yet other body modifications. "For most people, if you get a car, you're going to mod that car," says Licina. "Because if you don't crack it, you don't open it up, then it's not *yours*." For people with even simple implants, those alterations truly do become part of the self. When I ask Graafstra why it matters that his chips are in his hand, not on a card, he echoes Haworth's feelings about his magnet. "Psychologically, it is different," says Graafstra. "This is always with me. It changes my perception of self." After a decade, he says, "I consider it to be a part of my personal capabilities now, like that's a part of me."

The people around this table point out that it's already acceptable to augment the body and mind through education, exercise, and ordinary devices like eyeglasses. And there's already a lucrative industry based on voluntary augments for healthy people: cosmetic surgery. Sure, you could contend that these merely push people to the extremes of human beauty, fitness, or intelligence norms, not beyond them. But as Wolbring pointed out before, our expectation of "normal" keeps rising. It was once normal to die when exposed to the measles virus, but the body augment called vaccination is now widely considered a social good. Average height and lifespan changed dramatically within centuries, thanks to sanitation and nutrition technologies. And of course the idea of set biological limits is a deeply fraught one, which has been used perniciously against women, religious groups, and people of color to deny them education, workplace opportunities, and human rights. Where, the grinders ask, do you draw the line at which people must stop improving themselves?

Right now, for the most invasive technologies, the line seems to stop at medical need. When I'd first talked to Grindhouse members, they'd had a "Where's my jet pack?" moment of griping over the nonavailability of robot parts on the consumer market. "If you have your arm blown off in a war," said Cannon, "you can totally have a robot arm. But if you just want one, yeah, you're out of luck."

"And you would be considered mentally ill!" added Dimoveo.

At the very least, you'd be making a bad bargain. Today's prosthetic arms aren't nearly as dexterous as your own. And getting one is horribly invasive, Wolbring points out, which is why he thinks that even if neuroprosthetics become more capable than natural body parts, mainstream buyers will choose wearable alternatives: You'll buy a robotic exoskeleton, not a prosthetic leg. Augments being designed for entertainment—not medical purposes—are all wearables like glasses, he points out. People won't want to deal with the risk of surgery, the price, the insurance, another surgery whenever there's an upgrade. It's not worth a brain-machine interface to play a video game, he says: "I mean, who wants to do that just to thought-control *Warcraft*?"

As with wearables, implanted technology brings up the perpetual arms race issue: rising expectations, pressure to conform, advantages accruing to the enhanced, perhaps even the creation of a techno-underclass, or the reinforcement of already-existing economic, racial, and gender divides. In theory, says Wolbring, he doesn't care if you give yourself ten legs. The problem is what happens if ten legs become seen as more productive than two: "So the ones who so far were totally fine with abilities based on two legs will be seen as just as deficient as the ones without any legs."

Grindhouse members are also grappling with this problem of exclusion. One evening, Dimoveo had been imagining how music or painting might change if artists could work with an expanded perceptual palette. "I think that we'll begin to see augments for enjoyment once art begins to center around these new senses," he said. "Then you're talking about a different way to reach people emotionally." But Cannon jumped in: "In a way, that begins to become exclusionary, in that a normal unaugmented human could stare at the canvas forever and never see the art."

Their solution—like most other biohackers'—is to make information open source, or publicly available and free to use or modify, so anyone can build their devices. When I mention the access question to the San Francisco group, the entire table cries, in joyful chorus, "*That's why it's open source!*" Yet they readily agree that open source doesn't necessarily mean cheap, or that everyone can—or will want to—build their own. Still, they think, it will put a crimp on monopolies and price gouging, and let you know what your technology really does and where your data goes.

That said, there are members of the modification community who favor *more* professionalization, and would like to work more openly with university and medical researchers who share their interests. One night as we are chatting, Samppa Von Cyborg takes out his phone to show me a coupling post for prosthetic eyes and fingers, not dissimilar to what he uses for transdermal jewelry. He'd like more cooperation between piercers and doctors, and better training for body modifiers, to minimize hackwork and accidents. He envisions a government-accredited cross between an art college and a medical school, where students could train under experienced modders and receive certificates allowing them to do procedures no deeper than the subdermis. "Every country has their art schools," he says. "So why not body art school?"

If it's hard to imagine a school for implants—well, it wasn't that long ago that piercing, and even tattooing, were considered pretty strange, even stigmatized. Von Cyborg can remember the exact moment when piercing went mainstream: 1993, the day Aerosmith released the music video for "Cryin," in which a young Alicia Silverstone gets a belly button ring. "That changed everything in one night," he recalls. "That was the first time when people actually saw something else than lobe piercings. So obviously it was kind of, *'Wow, is this possible?'*" Soon, he was piercing 30 navels a day. Now your local cheer squad captain probably has a belly ring. And as what once seemed edgy became vanilla, the field expanded to look for a new edge. Today Von Cyborg is working on RFID chips, motorized implants that vibrate, and LEDs that pulse in time with music or your heartbeat.

At Grindhouse, they're cautious about licenses or tuition because they're perpetually leery of financial barriers, but love the ideas of body art schools and community standards for good work. Ultimately, they would like to see their fledgling industry move from mail order and piercing studios to commercial manufacturing and storefront clinics, to see their products become as common as the belly ring. As Cannon puts it: "We can't wait for a world where you walk in and say 'One Circadia and one Southpaw, please, and I'm thinking about replacing this eye. What do you have in a turquoise?'"

IT'S LATE AT NIGHT, and the Grindhouse members are in their basement, working on the Northstar. At this point, it's mostly wires on a breadboard. They've envisioned two designs: a full version with the in-hand light-up compass, which would brighten as the wearer faces north, and a lightweight version, one with more immediate consumer appeal, they think, that would simply light up, a pretty hacker toy.

Cannon is camped out at a coffee table with a zillion electronic bits spread out next to a black laptop. Sarver perches over a soldering iron at his workbench, a double jeweler's loupe clipped to his glasses for extra magnification. Dimoveo, who has been roasting vegetables for a curry, comes downstairs and works quietly on his computer. This project is at its very beginning, so tonight it's just basics: trying to build a microcontroller that can log data and light an LED using very low power, necessary for a long-term implant. Their goal is to have it cycle through eight seconds of sleep, then wake for the ninth, so each battery charge will last a few weeks.

Progress moves slowly, especially when you are working with basically zero dollars. So over the next three days, they hash through these tasks: Getting their low-power board to log data to a memory card. Programming a three-color LED to cycle between red, blue, and green. Testing the sleep timer and wake function using a nightingale schematic, which makes a cheerful chirping noise to indicate it's gone into sleep mode. Ideally, all of these would work at the same time. But there's always something. The stray bit of solder that shorts out a connection. The LED that is in backward, or is doing the bad blink that indicates a coding problem. The log files that turn up empty. The thing that stops working after it worked fine before. The thing that starts working the minute they give up on it.

"Tedium abounds," says Cannon drily after the umpteenth LED problem. It doesn't really matter. They are perfectly content. Sarver answers all of Cannon's requests, no matter how impossible, with a steady "All righty." The atmosphere is sort of like a slumber party, but with more soldering. The banter is nonstop and the jokes extremely inside. Grindhouse members often address each other in a fake courtly tone, ending everything with "sir," making them sound like they are constantly on the verge of a duel. (Sarver, staring at a breadboard on which nothing is, infuriatingly, happening: "Something is fishy in Denmark, sir.")

They're happy even when the hours get late, and one thousand cigarettes have been stubbed out in a kombucha bottle, and the rest of the house goes to bed or out for the night and it's just Sarver trying to write code to the memory card. Does he ever ask himself, you know, what he is doing in this unheated basement on a Saturday night? "No," he says. "Because it would be another unheated basement somewhere else."

And three nights later, they've got things working. Sort of. As they squint at the circuit, I run past them a working hypothesis for how magnetic implants function. Now, there's not exactly a Department of Magnetic Fingers that you can dial to ask an expert. But between Grindhouse visits I have been out to see neuroscientist Daniel Wesson at Case Western, who in addition to being an olfaction expert teaches a course on animal sensory perception, and is intrigued by the magnet idea.

First, he'd quashed the idea that implanting magnets gives you a unique modality. "They are never going to be able to experience anything beyond our five senses. We don't have the brain circuits for that," Wesson said. There's just no way you can stimulate the brain through the peripheral nervous system without using one of the five already-existing sensory channels. If you're putting magnets in contact with the hand's nerve fibers, he said, "The hand is designed to sense touch. You are only going to sense touch, whether the touch comes from a hot pan or from a magnet displacing the skin."

That said, Wesson pointed out excitedly, you can give the touch channel a novel input, and the brain, because of its extraordinary plasticity, learns to make patterns from that noise. The magnet reacts to its environment, it moves, you feel that pressure through touch, and eventually, with enough repeated exposure, your brain learns a new pattern. This, in short, is sensory substitution, and it's been used to create assistive technologies, like devices that convert images into tactile cues to help the blind navigate. "The whole purpose of our nerve system is to transduce environmental energy into meaningful codes," Wesson said. "What [magnets] are doing is allowing us to detect a new type of environmental energy." So it's not a new sense; it's new information being ported through an old sense. It's touch, *plus*.

The other scientists I run this by independently come to the same conclusion. Neuroscientist Dean Buonomano from UCLA, the time expert who is used to studying a perceptual experience with no associated organ, agrees the information is being relayed through touch. He compares a magnet implant to using a Geiger counter, which transforms radiation (which you cannot ordinarily sense) into sound waves (which you can). And whether you can call this a sixth sense, Buonomano says, depends on how you define your terms. If a sixth sense is a new part of the brain that senses magnetism, "then obviously the answer is no." But if you define it "as an ability to pick up physical stimuli from an environment that we wouldn't normally pick up, then the answer is absolutely 100 percent yes," he says.

"These people are definitely not getting a magnetic sense in the truest way, but I suppose it's possible that they are getting enough of a touch sense that they can do something with it," adds Sonke Johnsen, the magnetoreception expert. "The ability of the neural system to adapt to new information and incorporate it is nothing short of incredible. That's the way Braille works."

When I test this idea out on the assorted biohackers of my acquaintance, it's met with a universal wave of relieved agreement. "That is exactly right!" exclaims Cannon. "That makes perfect sense," agrees Sarver. "I feel the electromagnetic spectrum through a sense of touch."

Here's Amal Graafstra's take: "The brain has learned to reinterpret the data coming in from the tactile sense. You are co-opting that pathway to the brain." And here, finally, is Todd Huffman, the only guy I know with a finger magnet *and* a neuroscience degree: "It's absolutely piggybacking on an existing sense," specifically touch, he says. And that may be as close to consensus as we're going to get without anyone opening a Department of Magnetic Fingers.

So with that, the Grindhouse guys get back to the Northstar, essentially another sensory substitution concept, since it would convey information about magnetic orientation back to the brain via sight. Earlier that evening they had the data logging working on low power. Now they're trying to add in the sleep program. But as they try to push new code onto the card, it locks up. So they try again. And it sticks. And it sticks again. And now the LED is back to doing the bad blink. "Now I'm really confused," says Cannon, poking at the circuit before sitting back resignedly. "All right. We'll have to go back to research on that."

"All righty," says Sarver cheerfully.

And that's when I turn off the tape recorder, close my notebook, and head back up the Grindhouse stairs. Because whatever is going to happen isn't going to happen tonight. It might not even happen by the time you read this. Trying to outpace evolution is incredibly hard. Science might move faster than random mutation, but evolution's got the power of parallel processing; it can run more simultaneous experiments than all the laboratories in the world. And if you're on a basement budget, you can only try so many variations at once. By the following year, the group had decided to work first on the lightweight version, and leave the compass element for later. But they had also revived their Bottlenose sonar design and were working on a DIY kit that would let people build a glove-based model. They'd become an LLC business. They were somehow simultaneously both underground and above it. Most of all, they were still grinding.

And the rest of the world kept grinding, too. In 2015, Apple released the Apple Watch, which along with its smartwatch competitors that offer limited biometric data collection struck me as a cousin of the Circadia, one you actually *could* buy at the mall. (Through its "Digital Touch" feature, it also lets wearers send tap patterns and even their heartbeats to the wrists of Apple Watch–wearing friends, which is straight-up *Doktor Sleepless*.) The Science for the Masses group had moved on from their dietary restriction experiment in pursuit of near infrared vision, and had instead tried dropping a chlorophyll analog straight into Gabriel Licina's eyeball in an effort to give him night vision. David Eagleman's group at Baylor announced they had designed a vest that would translate information from the stock market or Twitter into tactile patterns a person could sense through vibrations on their back. A group at Duke University had already revealed a brain-controlled exoskeleton piloted by a parapalegic man; the machine translated feedback from the robot's movement and contact with the ground into tiny vibrators in a "smart shirt" so the man could feel as though he were walking under his own power. The BrainGate team at Brown University announced they had developed a brain-machine interface that's *wireless*. Almost certainly, somewhere a biohacker we've never heard of did something we can't yet imagine. The cutting edge moved, as the edge always does.

The quest to enhance or extend sensory perception is, in a strange way, an exploration of the limits of our limits. As long as we're still outfitted with the standard-issue human brain, we'll only have five ways to port information in. That is our fundamental perceptual limitation. But within that boundary lies the tantalizing possibility of clever workarounds, ways to play with the processing of those five data streams. So far, most have been attempts to expand the ranges of the five modalities, or to develop technologies that can restore their function, or to jerry-rig a "sixth" by funneling new kinds of information through a channel operated by the original five. Some of these, like magnets and retinal implants, are teasers from the sci-fi future, a glimpse of a world in which sensory devices under the skin can intelligibly talk to the brain. And some of them, like the search for the sixth taste, are much gentler: the idea that we can alter perception by learning new ways to categorize and pay attention to the information stream in which we already exist.

Ultimately, these are all hacks upon the already hacked. Neuroscience shows us that the brain is a filtering machine, which warps our perception for our own good, making sensory input seem synchronous, linear, complete, even when it is not. Social science shows us that we constantly shape our attention through language, experience, and cultural practices, creating a social world in which we can meaningfully communicate our inner states to others. And everything we know about the brain tells us that it is a voracious assimilator of new information; it will learn to read whatever we can learn to write in. We are both the hackers and the hacked, the modified and the modifier, the reader and the writer. Nature is amazing, say the biohackers, and before them the engineers, the doctors, the athletes, the teachers—anyone who has ever called upon us to transform our minds and bodies through our own efforts. *Couldn't it be more so?*

It is a beautiful, frustrating, wild proposition. The universe of information is so big, and our reality is so small. We know there are things we do not know; we struggle to imagine what we cannot feel. It is heartbreakingly human to want to be more than human, creatures that can not just *do* more, but *experience* more. And so we keep building toward the edge, wherever it is. "How we perceive the world is based on our current hardware, which was the best that accident had to offer," Tim Cannon once told me. "It's pretty good for lightning and mud. But I think we can have a better go at it."

Acknowledgments

This book was researched through the awesome power of sofa surfing. Thank you to everyone who put me up and put up with me. You made this book financially possible and incredibly fun to write: Laura Killips (Philadelphia), Jacquie Brown (Toronto), Jessica Wurster (Montréal), the Karesh family and the Maheta-Roig family (Washington, D.C., area), the extended Platoni clan (Southern California and New Jersey), Melecio Flores (Phoenix), the Maxwell family (Denver), Shannon Service (secret writing hideout), Stefania Rousselle (Paris), and the Laurison family (London).

I'd like to extend a special thank-you to everyone in this book who invited me into their office, lab, or home for a round of 3,000 Questions, or who emailed or Skyped their thoughts to me from across vast distances. I am grateful for your patience, your insights, and your incredible work exploring the field of perception. My particular thanks to Daniel Wesson, Michael Tordoff, Nicole Garneau, Rachel Herz, and Krishna Shenoy, who were especially generous with their time and fact-checking help. Thank you also to the people who don't appear in these pages but whose kindness and savvy led me to those who do, especially Ken Goldberg, Chris Loss, Matthias Tabert, Sandor Katz, Peter Van Tassel, Sam Rolens, Jason Jaacks, and Dylan Bergeson, captain of the drone strikes.

And finally, thank you to my brothers in arms Eric Simons and Jonathan Kauffman, without whom I never would have written a book and would have been lonely doing it; to my friend and mentor Cynthia Gorney, who taught me everything I know (twice); to super agent Gillian MacKenzie (who sparked this book's idea) and to my wonderful editors Tisse Takagi (who saw its promise) and Alison Mackeen (who saw it through); to my great friends Casey Miner, Zoe Gladstone, Lynn Deregowski, and Jenny Maxwell for every kindness and soothing word; to the UC Berkeley Graduate School of Journalism and my students and colleagues who cheered me on and kept me caffeinated; to Mike Smith and Leah Platoni for the bedrock of home team support; and to my parents, Bob and Alexis Platoni, who taught me to love books in the first place. I'm sorry, guys, for all the years I spent reading by the nightlight.

Notes

Chapter 1

4 *amino acid glutamate*: Kikunae Ikeda, "New Seasonings," *Journal of the Chemical Society of Tokyo* 30 (1909): 820–836.

4 *one of the leading contenders*: Richard Mattes, "Is There a Fatty Acid Taste?" *Annual Review of Nutrition* 29 (2009): 305–327.

4 *rules of thumb*: Robin Tucker and Richard Mattes, "Are Free Fatty Acids Effective Taste Stimulus in Humans?" (paper presented at the Institute of Food Technologists 2011 Annual Meting, New Orleans, Louisiana, June 12, 2011).

5 *coconut and almonds*: Bhushan Kulkarni and Richard Mattes, "Evidence for Presence of Nonesterified Fatty Acids as Potential Gustatory Signaling Molecules in Humans," *Chemical Senses* 38, no. 2 (2012): 119–127.

10 *some preliminary results*: Robin Tucker et al., "No Difference in Perceived Intensity of Linoleic Acid in the Oral Cavity Between Obese and Non-obese Adults" (poster presented at Experimental Biology, Boston, Massachusetts, March 30, 2015).

12 *flawed organizational structure*: Jeannine Delwiche, "Are There 'Basic' Tastes?" *Trends in Food Science and Technology* 7 (1996): 411–415.

14 *the T1R3 receptor*: Michael Tordoff et al., "Involvement of T1R3 in Calcium-Magnesium Taste," *Physiological Genomics* 34 (2008): 338–348.

15 *dosed with lactisole*: Michael Tordoff et al., "T1R3: A Human Calcium Taste Receptor," *Scientific Reports* 2, no. 496 (2012), doi:10.1038/srep00496.

16 *taste carbon dioxide*: Charles Zuker et al., "The Taste of Carbonation," *Science* 326 (2009): 443–445.

16 *three carbohydrate tastes:* Anthony Sclafani, "The Sixth Taste?" *Appetite* 43 (2004): 1–3.

17 *calcium-sensing receptor, CaSR*: Yuzuru Eto et al., "Kokumi Substances, Enhancers of Basic Tastes, Induce Responses in Calcium-Sensing Receptor Expressing Taste Cells," *PLoS ONE* 7, no. 4 (2012), doi:10.1371/journal.pone.0034489.

17 *taste foods laced with these kokumi-related substances*: Yuzuru Eto et al., "Involvement of Calcium-Sensing Receptor in Human Taste Perception," *Journal of Biological Chemistry* 285, no. 2 (2010): 1016–1022.

17 *commercial soy and fish sauces*: Yuzuru Eto et al., "Determination and Quantification of the Kokumi Peptide, c-glutamyl-valyl-glycine, in Commercial Soy Sauces," *Food Chemistry* 141 (2013): 823–828.

Chapter 2

30 *such as Parkinson's*: Richard Doty, "Olfaction in Parkinson's Disease and Related Disorders," *Neurobiology of Disease* 46, no. 3 (2012): 527–552.

33 *47.5 million dementia patients*: World Health Organization, "Dementia Fact Sheet," March 2015, www.who.int/mediacentre/factsheets/fs362/en/.

37 *over 1 trillion*: Andreas Keller et al., "Humans Can Discriminate More than 1 Trillion Olfactory Stimuli," *Science* 343, no. 6177 (2014): 1370–1372.

37 *The anatomic positioning*: Rachel Herz, *The Scent of Desire: Discovering Our Enigmatic Sense of Smell* (New York: HarperCollins, 2007), 3–6, 13–24, 50–52, 63–73, 238.

38 *crayons, Play-Doh and Coppertone*: Rachel Herz and Jonathan Schooler, "A Naturalistic Study of Autobiographical Memories Evoked by Olfactory and Visual Cues: Testing the Proustian Hypothesis," *American Journal of Psychology* 115, no. 1, (2002): 21–32.

38 *freshly cut grass*: Rachel Herz, "A Naturalistic Analysis of Autobiographical Memories Triggered by Olfactory Visual and Auditory Stimuli," *Chemical Senses* 29 (2004): 217–224.

39 *first appearance of Alzheimer's pathogens*: Heiko Braak and Eva Braak, "Neuropathological Staging of Alzheimer-Related Changes," *Acta Neuropathologica* 82 (1991): 239–259.

39 *the staging of Parkinson's*: Heiko Braak et al., "Staging of Brain Pathology Related to Sporadic Parkinson's Disease," *Neurobiology of Aging* 24, no. 2 (2003): 197–211.

42 *correlated with the spread of amyloid*: Daniel Wesson et al., "Olfactory Dysfunction Correlates with Amyloid-β Burden in an Alzheimer's Disease Mouse Model," *Journal of Neuroscience* 30, no. 2 (2010): 505–514.

42 *habituating to odors*: Wen Li, James Howard, and Jay Gottfried, "Disruption of Odour Quality Coding in Piriform Cortex Mediates Olfactory Deficits in Alzheimer's Disease," *Brain* 133 (2010): 2714–2726.

42 *increasingly abnormal*: Daniel Wesson et al., "Sensory Network Dysfunction, Behavioral Impairments, and Their Reversibility in an Alzheimer's β-Amyloidosis Mouse Model," *Journal of Neuroscience* 31, no. 44 (2011): 15962–15971.

46 *lead to better diagnoses*: Davangere Devanand et al., "Combining Early Markers Strongly Predicts Conversion from Mild Cognitive Impairment to Alzheimer's Disease," *Biological Psychiatry* 64, no. 10 (2008): 871–879.

49 *Taiwan, Australia, and Brazil*: Marco Fornazieri et al., "A New Cultural Adaptation of the University of Pennsylvania Smell Identification Test," *CLINICS* 68, no. 1 (2013): 65–68.

Chapter 3

60 *earlier sensory prosthetic*: National Institute on Deafness and Other Hearing Disorders, "Cochlear Implants," last updated August 2014, www.nidcd.nih.gov/health/hearing/pages/coch.aspx.

60 *Researchers began experimenting*: Gretchen Henkel, "History of the Cochlear Implant," *ENT Today*, April 2013.

61 *probes within the eye*: Mark Humayun et al., "Visual Perception Elicited by Electrical Stimulation of Retina in Blind Humans," *Archives of Ophthalmology* 114 (1996): 40–46; Mark Humayun et al., "Pattern Electrical Stimulation of the Human Retina," *Vision Research* 39 (1999): 2569–2576.

66 *Retina Implant AG*: Eberhart Zrenner et al., "Artificial Vision with Wirelessly Powered Subretinal Electronic Implant Alpha-IMS," *Proceedings of the Royal Society B* 280 (2013), doi:10.1098/rspb.2013.0077.

66 *envisions a system*: Sheila Nirenberg and Chethan Pandarinath, "Retinal Prosthetic Strategy with the Capacity to Restore Normal Vision," *Proceedings of the National Academy of Sciences* 109, no. 37 (2012), doi:10.1073/pnas.1207035109.

66 *close enough to normal input*: Sheila Nirenberg, "A Prosthetic Eye to Treat Blindness," TEDMED Talk, October 2011, http://www.ted.com/talks/sheila_nirenberg_a_prosthetic _eye_to_treat_blindness.

Chapter 4

75 *Several labs at UC Berkeley*: Full disclosure: I teach at UC Berkeley in the Graduate School of Journalism, but was not employed there during the researching of this book.

78 *categorical perception*: Edward Chang et al., "Categorical Speech Representation in Human Superior Temporal Gyrus," *Nature Neuroscience* 13, no. 11 (2010): 1428–1432.

80 *listened to spoken phrases*: Edward Chang et al., "Functional Organization of Human Sensorimotor Cortex for Speech Articulation," *Nature* 495 (2013): 327–332.

81 *two different decoding strategies*: Brian Pasley et al., "Reconstructing Speech from Human Auditory Cortex," *PLoS Biology* 10, no. 1 (2012), doi:10.1371/journal.pbio.1001251.

83 *reconstruction of covert speech*: Brian Pasley et al., "Decoding Spectrotemporal Features of Overt and Covert Speech from the Human Cortex," *Frontiers in Neuroengineering* 7, no. 14 (2014), doi:10.3389/fneng.2014.00014.

86 *using still images*: Jack Gallant et al., "Identifying Natural Images from Human Brain Activity," *Nature* 452 (2008): 352–355.

86 *subjects looked at photographs*: Jack Gallant et al., "Bayesian Reconstruction of Natural Images from Human Brain Activity," *Neuron* 63 (2009): 902–915.

87 *showing them videos*: Jack Gallant et al., "Reconstructing Visual Experiences from Brain Activity," *Current Biology* 21 (2011): 1641–1646.

89 *continuous semantic space*: Alexander Huth et al., "A Continuous Semantic Space Describes the Representation of Thousands of Object and Action Categories across the Human Brain," *Neuron* 76, (2012): 1210–1224.

90 *reconstructing remembered images*: Thomas Naselaris et al., "A Voxel-wise Encoding Model for Early Visual Areas Decodes Mental Images of Remembered Scenes," *NeuroImage* 105 (2014): 215–228.

91 *start with dreams*: Tomoyasu Horikawa et al., "Neural Decoding of Visual Imagery during Sleep," *Science* 340 (2013): 639–642.

Chapter 5

96 *sensory substitution*: Allison Okamura et al., "Force Feedback and Sensory Substitution for Robot-assisted Surgery," in *Surgical Robotics: Systems, Applications and Visions*, ed. Jacob Rosen, Blake Hannaford, and Richard Satava (New York: Springer, 2010), 419–448.

98 *skin stretch*: Zhan Fan Quek et al., "Sensory Augmentation of Stiffness Using Fingerpad Skin Stretch," *IEEE World Haptics Conference* (2013): 467–472.

100 *haptic jamming*: Andrew Stanley and Allison Okamura, "Controllable Surface Haptics via Particle Jamming and Pneumatics," *IEEE Transactions on Haptics* 8, no. 1 (2015): 20–30.

101 *virtual reality display*: Andrew Stanley et al., "Integration of a Particle Jamming Tactile Display with a Cable-driven Parallel Robot," in *Haptics: Neuroscience, Devices,*

Modeling, and Applications; Proceedings of the Eurohaptics Conference (New York: Springer, 2014), 258–265.

107 *transoceanic distances*: Jacques Marescaux et al., "Transcontinental Robot-assisted Remote Telesurgery: Feasibility and Potential Applications," *Annals of Surgery* 235, no. 4 (2002): 487–492.

108 *telerobotics surgical service*: Mehran Anvari, Craig McKinley, and Harvey Stein, "Establishment of the World's First Telerobotic Remote Surgical Service," *Annals of Surgery* 241 (2005): 460–464.

108 *leaders in neuroprosthetics research*: Krishna Shenoy et al., "Challenges and Opportunities for Next-generation Intracortically Based Neural Prostheses," *IEEE Transactions on Biomedical Engineering* 58, no. 7 (2011): 1891–1899.

Chapter 6

120 *lack of a central counter*: Uma Karmarkar and Dean Buonomano, "Timing in the Absence of Clocks: Encoding Time in Neural Network States," *Neuron* 53 (2007): 427–438.

121 *striatal-beat frequency model*: Warren Meck, Trevor Penney, and Viviane Pouthas, "Cortico-striatal Representation of Time in Animals and Humans," *Current Opinion in Neurobiology* 18 (2008): 145–152.

121 *cerebellum regulates timing*: Richard Ivry and Rebecca Spencer, "The Neural Representation of Time," *Current Opinion in Neurobiology* 14 (2004): 225–232.

121 *tells time through neural dynamics*: Dean Buonomano and Rodrigo Laje, "Population Clocks: Motor Timing with Neural Dynamics," *Trends in Cognitive Sciences* 14, no. 12 (2010): 520–527.

123 *switch order*: David Eagleman, "Human Time Perception and Its Illusions," *Current Opinion in Neurobiology* 18 (2008): 131–136.

123 *whether fear actually "slows" time*: Chess Stetson et al., "Does Time Really Slow Down during a Frightening Event?" *PLoS ONE* 2, no. 12 (2007), doi:10.1371/journal.pone .0001295.

123 *slowest-arriving sensory feed*: David Eagleman, "Brain Time," in *What's Next? Dispatches from the Future of Science*, ed. Max Brockman (New York: Vintage, 2009), http://edge.org/conversation/brain-time.

131 *teach rat brain slices to tell time*: Hope Johnson, Anubhuthi Goel, and Dean Buonomano, "Neural Dynamics of in vitro Cortical Networks Reflects Experienced Temporal Patterns," *Nature Neuroscience* 13, no. 8 (2010): 917–919.

Chapter 7

144 *rejecting Cyberball players*: Naomi Eisenberger, Matthew Lieberman, and Kipling Williams, "Does Rejection Hurt? An fMRI Study of Social Exclusion," *Science* 302 (2003): 290–292.

144 *variations on this theme*: Naomi Eisenberger, "Broken Hearts and Broken Bones: A Neural Perspective on the Similarities between Social and Physical Pain," *Current Directions in Psychological Science* 21, no. 1 (2012): 42–47.

144 *unwanted breakups*: Ethan Kross et al., "Social Rejection Shares Somatosensory Representations with Physical Pain," *Proceedings of the National Academy of Sciences* 108, no. 15 (2011): 6270–6275.

144 *pain-killing power of Tylenol*: C. Nathan DeWall et al., "Acetaminophen Reduces So-cial Pain: Behavioral and Neural Evidence," *Psychological Science* 21, no. 7 (2010): 931–937.

145 *effect of pot on social pain*: Timothy Deckman et al., "Can Marijuana Reduce Social Pain?" *Social Psychological and Personality Science* 5, no. 2 (2013), doi:10.1177/1948550613488949.

148 *reliving pain*: Meghan Meyer, Kipling Williams, and Naomi Eisenberger, "Why Social Pain Can Live On: Different Neural Mechanisms Are Associated with Reliving Social and Physical Pain," *PLoS ONE* 10, no. 6. (2015): doi:10.1371/journal.pone.0128294.

154 *with math*: Ian Lyons and Sian Beilock, "When Math Hurts: Math Anxiety Predicts Pain Network Activation in Anticipation of Doing Math," *PLoS ONE* 7, no. 10 (2012), doi:10.1371/journal.pone.0048076.

155 *passionate love*: Jarred Younger et al., "Viewing Pictures of a Romantic Partner Re-duces Experimental Pain: Involvement of Neural Reward Systems," *PLoS ONE* 5, no. 10 (2010), doi:10.1371/journal.pone.0013309.

156 *established couples*: Sarah Master et al., "Partner Photographs Reduce Experimen-tally Induced Pain," *Psychological Science* 20, no. 11 (2009): 1316–1318.

156 *diminish your pain response*: Naomi Eisenberger et al., "Attachment Figures Activate a Safety Signal–Related Neural Region and Reduce Pain Experience," *Proceedings of the National Academy of Sciences* 108, no. 28 (2011): 11721–11726.

157 *makes you feel less threat*: Tristen Inakagi and Naomi Eisenberger, "Neural Correlates of Giving Support to a Loved One," *Psychosomatic Medicine* 74 (2012): 3–7.

Chapter 8

160 *culture directs how you experience emotion*: Yulia Chentsova-Dutton, Andrew Ryder, and Jeanne Tsai, "Understanding Depression across Cultural Contexts," in *Handbook of Depression*, 3rd ed., ed. Ian Gotlib and Constance Hammen (New York: Guilford, 2014), 337–354.

160 *processing sadness or depression*: Andrew Ryder and Yulia Chentsova-Dutton, "Depres-sion in Cultural Context: 'Chinese Somatization,' Revisited," *Psychiatric Clinics of North America* 35 (2012): 15–36.

163 *limbic system*: Paul MacLean, "Psychosomatic Disease and the 'Visceral Brain,' " *Psy-chosomatic Medicine* 11, no. 6 (1949): 338–352.

165 *neck pain can trigger panic attacks*: Devon Hinton and Michael Otto, "Symptom Pre-sentation and Symptom Meaning among Traumatized Cambodian Refugees: Relevance to a Somatically Focused Cognitive-Behavior Therapy," *Cognitive and Behavioral Practice* 13, no. 4 (2009): 249–260.

168 *Participant 57 was an outlier*: Yulia Chentsova-Dutton et al., "Chinese Americans Report More Somatic Experiences than European Americans Following a Sad Film" (poster presented at Association for Psychological Science Annual Convention, San Francisco, Cali-fornia, May 22–25, 2014).

169 *what you want to feel*: Jeanne Tsai, "Ideal Affect: Cultural Causes and Behavioral Con-sequences," *Perspectives on Psychological Science* 2, no. 3 (2007): 242–259.

169 *children's storybooks*: Jeanne Tsai et al., "Learning What Feelings to Desire: Social-ization of Ideal Affect through Children's Storybooks," *Personality and Social Psychology Bul-letin* 33, no. 17 (2007): 17–30.

170 *defy their culture's ideal affect*: Yulia Chentsova-Dutton et al., "Depression and Emo-tional Reactivity: Variation among Asian Americans of East Asian Descent and European Americans," *Journal of Abnormal Psychology* 116, no. 4 (2002): 776–785.

171 *assistance to being shunned*: Andrew Ryder et al., "The Cultural Shaping of Depression: Somatic Symptoms in China, Psychological Symptoms in North America?" *Journal of Abnormal Psychology* 117, no. 2 (2008): 300–313.

172 *situation that made them angry*: Eunsoo Choi and Yulia Chentsova-Dutton, "Distress Experience and Expression in Cultural Contexts: Examination of Koreans and Americans" (poster presented at the Annual Meeting of the Society of Personality and Social Psychology, Austin, Texas, February 13–15, 2014).

175 *but there was a twist*: Biru Zhou et al., "Ask and You Shall Receive: Actor-Partner Interdependence Model Approach to Estimate Cultural and Gender Variations in Social Support Seeking and Provision Behaviours," (2015), unpublished manuscript.

176 *compared politicians' smiles*: Jeanne Tsai et al., "Leaders' Smiles Reflect Their Nations' Ideal Affect," (2014), unpublished manuscript.

177 *Christian texts and worshippers*: Jeanne Tsai, Felicity Miao, and Emma Seppala, "Good Feelings in Christianity and Buddhism: Religious Differences in Ideal Affect," *Personality and Social Psychology Bulletin* 33, no. 409 (2007): 409–421.

177 *choose a doctor*: Tamara Sims et al., "Choosing a Physician Depends on How You Want to Feel," *Emotion* 14, no. 1 (2014), 187–192.

177 *virtual doctor*: Tamara Sims and Jeanne Tsai, "Patients Respond More Positively to Physicians Who Focus on Their Ideal Affect," *Emotion* 15, no. 3 (2014), 303–318.

178 *affective or emotional mechanism*: BoKyung Park et al., "Neural Evidence for Cultural Differences in the Valuation of Positive Facial Expressions," (2015), unpublished manuscript.

Chapter 9

184 *problem for the military*: Hannah Fischer, "A Guide to U.S. Military Casualty Statistics," Congressional Research Service report, February 19, 2014, www.crs.gov.

185 *the STRIVE program*: Albert Rizzo et al., "Virtual Reality as a Tool for Delivering PTSD Exposure Therapy and Stress Resilience Training," *Military Behavioral Health* 1 (2013): 48–54.

188 *fear of heights*: Larry F. Hodges et al., "Virtual Environments for Treating the Fear of Heights," *IEEE Computer* 28, no. 7 (1995): 27–34.

189 *rolled out Virtual Vietnam*: Larry F. Hodges et al., "Virtual Vietnam: A Virtual Environment for the Treatment of Vietnam War Veterans with Post-traumatic Stress Disorder" (paper presented at the International Conference on Artificial Reality and Telexistence, Tokyo, Japan, 1998).

190 *the therapy helped*: Barbara Olasov Rothbaum et al., "Virtual Reality Exposure Therapy for PTSD Vietnam Veterans: A Case Study," *Journal of Traumatic Stress* 12, no. 2 (1999): 263–271.

190 *11 percent of army soldiers*: Charles Hoge et al., "Combat Duty in Iraq and Afghanistan, Mental Health Problems, and Barriers to Care," *New England Journal of Medicine* 351, no. 1 (2004): 13–22.

190 *effective for war veterans*: Albert Rizzo et al., "Virtual Reality Applications to Address the Wounds of War," *Psychiatric Annals* 43, no. 3 (2013): 123–138.

193 *carry over into real life*: Nick Yee and Jeremy Bailenson, "The Proteus Effect: The Effect of Transformed Self-Representation on Behavior," *Human Communication Research* 33 (2007): 271–290.

193 *real-life bargaining task*: Nick Yee and Jeremy Bailenson, "The Difference between Being and Seeing," *Media Psychology* 12 (2009): 195–209.

193 *exercise level*: Jesse Fox and Jeremy Bailenson, "Virtual Self-Modeling: The Effects of Vicarious Reinforcement and Identification on Exercise Behaviors," *Media Psychology* 12 (2009): 1–25.

193 *rape myth acceptance*: Jesse Fox, Jeremy Bailenson, and Liz Tricase, "The Embodiment of Sexualized Virtual Selves: The Proteus Effect and Experiences of Self-Objectification via Avatars," *Computers in Human Behavior* 29 (2013): 930–938.

195 *conserve energy*: Jakki Bailey et al., "The Impact of Vivid Messages on Reducing Energy Consumption Related to Hot Water Use," *Environment and Behavior* 17 (2015): 570–592.

196 *used fewer napkins*: Sun Joo (Grace) Ahn, Jeremy Bailenson, and Dooyeon Park, "Short- and Long-Term Effects of Embodied Experiences in Immersive Virtual Environments on Environmental Locus of Control and Behavior," *Computers in Human Behavior* 39 (2014): 235–245.

202 *adapt within five minutes*: Andrea Stevenson Won, "Homuncular Flexibility in Virtual Reality," *Journal of Computer Mediated Communication* 20, no. 3 (2015): 241–259.

Chapter 10

204 *inactive by default*: "FAQ," Google "Glass Press," https://sites.google.com/site/glasscomms/faqs.

207 *first Torontonian with some kind of eyecam*: Steve Mann, "My Augmediated Life," *IEEE Spectrum*, March 1, 2013, http://spectrum.ieee.org/geek-life/profiles/steve-mann-my-augmediated-life.

207 *Mann was an early collaborator*: "Implantable Camera System," http://wearcam.org/eyeborg.htm.

207 *Neil Harbisson*: Neil Harbisson, "I Listen to Color," TEDGlobal Talk, June 2012, http://www.ted.com/talks/neil_harbisson_i_listen_to_color.

207 *Waafa Bilal*: "3rdi," "About," www.3rdi.me.

207 *Thad Starner*: Jesse Lichtenstein, "Magnifying Glass," *Atlanta Magazine*, March 2, 2014, http://www.atlantamagazine.com/great-reads/magnifying-glass-thad-starner-google-glass/.

207 *2011 minidocumentary*: "Deus Ex: Human Revolution—The Eyeborg Documentary," August 2, 2011, http://eyeborgproject.com.

213 *coined the term*: Manfred Clynes and Nathan Kline, "Cyborgs and Space," *Astronautics*, September 1960, 26–27, 74–76.

214 *we are all cyborgs*: Donna Haraway, "Manifesto for Cyborgs: Science, Technology, and Socialist Feminism in the 1980s," *Socialist Review* 80 (1985): 65–108.

214 *"the world's first cyborg"*: Kevin Warwick, *I, Cyborg* (Chicago: University of Illinois Press, 2004), 61, 232–235, 260–264, 282–289.

216 *annoyed patrons or managers*: "Woman Wearing Google Glass Says She Was Attacked in San Francisco Bar," CBS, February 25, 2015, http://sanfrancisco.cbslocal.com/2014/02/25/woman-wearing-google-glass-says-she-was-attacked-in-san-francisco-bar/.

216 *etiquette guide*: "Explorers," Google, https://sites.google.com/site/glasscomms/glass-explorers.

216 *police officers tried to pull his camera off*: Neil Harbisson, "The Man Who Hears Colour," BBC, February 15, 2012, www.bbc.com/news/magazine-16681630.

216 *explaining their function*: "Physical Assault by McDonald's for Wearing Digital Eye Glass," *Steve Mann's Blog*, July 16, 2012, http://eyetap.blogspot.com/2012/07/physical-assault-by-mcdonalds-for.html.

217 *suspected him of bootlegging*: Emma Woollacott, "Homeland Security Hauls Man from Movie Theater for Wearing Google Glass," *Forbes*, January 22, 2014, www.forbes.com/sites /emmawoollacott/2014/01/22/homeland-security-hauls-man-from-movie-theater-for -wearing-google-glass/.

219 *mayor Rob Ford*: Daniel Dale, "Rob Ford: Yes, I Have Smoked Crack Cocaine," *Toronto Star*, November 5, 2013, http://www.thestar.com/news/crime/2013/11/05/rob_ford_yes _i_have_smoked_crack_cocaine.html.

219 *watching from "below"*: Steve Mann and Joseph Ferenbok, "New Media and the Power Politics of Sousveillance in a Surveillance-Dominated World," *Surveillance & Society* 11, no. 1/2 (2013): 18–34.

219 *false accusations*: Chris Gayomeli, "4 Innocent People Wrongly Accused of Being Boston Marathon Bombing Suspects," *The Week*, April 19, 2013, http://theweek.com/article /index/243028/4-innocent-people-wrongly-accused-of-being-boston-marathon-bombing -suspects.

221 *a bedtime squeeze*: James Teh and Adrian David Cheok, "Pet Internet and Huggy Pajama: A Comparative Analysis of Design Issues," *International Journal of Virtual Reality* 7, no. 4 (2008): 41–46.

221 *touch influences feelings*: Gilang Andi Pradana et al., "Emotional Priming of Mobile Text Messages with Ring-Shaped Wearable Device Using Color Lighting and Tactile Expressions" (paper presented at the Augmented Human International Conference, Kobe, Japan, March 7–9, 2014).

222 *conveys a kiss by phone*: Elaham Saadatian et al., "Mediating Intimacy in Long-Distance Relationships Using Kiss Messaging," *International Journal of Human-Computer Studies* 72, no. 10–11 (2014): 736–746.

223 *long-distance touch using air*: "Aireal: Interactive Tactile Experiences in Free Air," Disney Research, www.disneyresearch.com/project/aireal/.

224 *electrically stimulating taste buds*: Adrian David Cheok et al., "Digital Taste for Remote Multisensory Interactions" (poster presented at User Interface Software and Technology Symposium, Santa Barbara, California, October 16–19, 2011).

226 *ability expectations*: Gregor Wolbring, "Ethical Theories and Discourses through an Ability Expectations and Ableism Lens," *Asian Bioethics Review* 4, no. 4 (2012): 293–309.

Chapter 11

231 *veteran in the body modification world*: Samppa Von Cyborg, "Body Mod," http:// voncyb.org/#bodymod/.

234 *theories of how magnetoreception works*: Sonke Johnsen and Kenneth Lohmann, "Magnetoreception in Animals," *Physics Today* 61, no. 3 (2008) 29–35; Kenneth Lohmann, "Magnetic-Field Perception," *Nature News & Views* 464, no. 22 (2010): 1140–1142.

235 *radical pair hypothesis*: Thorsten Ritz, Salih Adem, and Klaus Schulten, "A Model for Photoreceptor-Based Magnetoreception in Birds," *Biophysical Journal* 78 (2000): 707–718.

236 *experimented with Monarch butterflies*: Patrick Guerra, Robert Gegear, and Steven Reppert, "A Magnetic Compass Aids Monarch Butterfly Migration," *Nature Communications* 5 (2014), doi:10.1038/ncomms5164.

236 *blindfolded students*: R. Robin Baker, "Goal Orientation by Blindfolded Humans after Long-Distance Displacement: Possible Involvement of a Magnetic Sense," *Science* 210 (1980): 555–557.

236 *magnetic material in their sinuses*: R. Robin Baker, "Sinal Magnetite and Direction Finding," *Physics & Technology* 15 (1984): 30–36.

237 *cryptochromes found in the human retina*: Lauren Foley, Robert Gegear, and Steven Reppert, "Human Cryptochrome Exhibits Light-Dependent Magnetosensitivity," *Nature Communications* 2 (2011), doi:10.1038/ncomms1364.

237 *going on at Grindhouse*: Grindhouse Wetware, "About Us," http://www.grind housewetware.com/.

239 *Lepht Anonym*: Lepht Anonym, "Sapiens Anonym," http://sapiensanonym.blogspot .com/.

242 *exploring magnetic implants*: Steve Haworth Modified, LLC, 2012, "Magnetic FAQ," http://stevehaworth.com/main/?page_id=871.

246 *"the Big Brother issue"*: Kevin Warwick, *I, Cyborg* (Chicago: University of Illinois Press, 2004), 82–89.

253 *Digital Touch*: "Apple Watch—New Ways to Connect," Apple, 2015, https://www .apple.com/watch/new-ways-to-connect/.

253 *stock market or Twitter*: David Eagleman, "Can We Create New Senses for Humans?" TED Talk, March 2015, http://www.ted.com/talks/david_eagleman_can_we_create_new _senses_for_humans?.

253 *group at Duke University*: Miguel Nicolelis, "Brain-to-Brain Communication Has Arrived. Here's How We Did It." TEDGlobal Talk, October 2014, https://www.ted.com/talks /miguel_nicolelis_brain_to_brain_communication_has_arrived_how_we_did_it.

254 *brain-machine interface that's wireless*: Antonio Regalado, "A Brain-Computer Interface that Works Wirelessly," *MIT Technology Review*, January 14, 2015, http://www.technology review.com/news/534206/a-brain-computer-interface-that-works-wirelessly.

Index

ability creep, 226–227
ability expectations, 226
acculturation, 20, 21
acetaminophen, as emotional pain relief, 144–145
addiction, passionate love and, 155–156
Ahn, Sun Joo (Grace), 195–196
AIREAL, 223–224
Ajinomoto, 16–17, 20, 23
alcohol, as pain relief, 140, 152
Alpha IMS, 66
Alzheimer's disease: cause of, 39–41; and loss of smell, 38–39, 41–44; projects related to, 30–32, 51–53; smell and testing for, 44–48; and smell's relationship to memories, 32–35
amygdala, 37, 157, 163
amyloid beta, 40, 41
Anonym, Lepht, 239, 240
anterior cingulate cortex, 139, 143
Antikythera mechanism, 135
Apple Watch, 253
Archambault, Marie-France, 33, 34
Archer, Mike, 3, 6, 8, 9
Argus I Retinal Prosthesis System, 61–62
Argus II Retinal Prosthesis System: development of, 61–62, 65–67; functioning of, 58–60; future possibilities for, 67–70; life and vision with, 54–57, 70–73
arousal, and categorization of emotion, 163
Ashford, Wes, 40–41
aspergillus, 23
Atelier Olfactif, 29–35, 48–51

atomic clock, 127–131
auditory imagery, 83
augmented reality (AR): concerns regarding, 215–220; current innovations in, 209–213; cyborgs and, 213–214; devices and possibilities for, 204–205; as evolution, 227–229; future of, 225–227; Rob Spence and, 203, 205–208; touch, taste, and smell in, 220–225
Augmented World Expo, 212
Avegant, 213

Babbage, Charles, 134
backward editing, 123
Bailenson, Jeremy, 193–194, 196–198, 200–201, 202
Bailey, Jakki, 195
Baker, Robin, 236
basic taste, 12–13, 16, 20–21
Beauchamp, Gary, 11–12, 13
behavior: augmented reality and, 215–216; virtual reality and influencing, 192–199
Bilal, Waafa, 207
biohacking and biohackers: concerns regarding, 245–249; defined, x; Grindhouse group's projects in, 237–242, 250–251, 253; institutionalization of, 249–250; and magnetoreception, 230–237, 243–245, 251–253. See also augmented reality (AR)
bitter receptors, 12, 13, 21
Blevins, Elizabeth, 175–176
bodily integrity, 248

body, emotion and, 164–167, 171–172
body modification. *See* biohacking and biohackers
body sovereignty, 248
Boston Marathon bombing, 219
Bottlenose, 240
Bouzari, Ali, 18–21, 25
Braak, Eva, 39
Braak, Heiko, 39
brain: and anatomic position of olfactory processing, 37; and augmented reality, 225; and experience of pain, 143; filtering ability of, xiii–xiv; implants and vision restoration, 69–70; and interpretation of social pain, 139; and neural activity for imagery versus sensory perception, 90; perception and, xii–xiii; and perception of others, 177–178; and perception of time, 119, 120–123, 131–133; and perceptual delay, 123–124; and process of hearing, 77–78; and recalling pain, 148; and relationship between social and physical pain, 143–144; speech and, 79–81; and stimulus reconstruction, 84–86; studying, xviii; and study of emotional pain, 140–141; study of hearing and, 74–77; visceral, 163
Brain-Gate2 Neural Interface System, 112
Brand, Stewart, 124
Braun, Marius, 221
breakups, painfulness of, 138–140, 146–147, 150–152
Brennan, Geraldine, 46
broken heart, painfulness of, 138–140, 146–147, 150–152
Buddhas of Bamiyan, 136
Buonomano, Dean, 120, 121–124, 131, 252
Burrus, Jim, 128, 130

calcium, tasting, 14–16
Camilli, Anne, 29–32, 33–35
Cannon, Tim, ix, x, 237–242, 248–254
caregiving, 157
Carlson, Kaitlin, 43
categorical perception, 78–79
cesium, 128–129, 130
Chang, Edward, 78–81
Chavez, Kenneth, 183–184, 186

Chentsova Dutton, Yulia, 159, 160–162, 164, 165–169, 170–171
Cheok, Adrian David, xix–xx, 220–223, 224–225, 227
Choi, Eunsoo, 171–172
Circadia, 237–239, 240
classification, and stimulus reconstruction, 85
clock(s): atomic, 127–131; of The Long Now Foundation, 117–119, 124–125, 133–137
Clynes, Manfred, 213–214
cochlear implants, 60–61
cognitive mechanism, 178
Colliau, Jennifer, 137
color, 57, 71–72
cones, 57–58
congenital insensitivity to pain, 153
Connaughton, Kyle, 18–21
Consumer Electronics Show, 212–213
coral, virtual reality as, 199–200
Cosmetic Executive Women, 29–30, 33
covert speech, 83–84
cow, virtual reality as, 198–199
Cray supercomputer, 134–135
cryptochromes, 235–236, 237
Cui, Dan, 211
culture: and adaptation to new tastes, 20, 21; and ideal affect, 169–170; and language and emotion, 164; and lexical challenge of describing unestablished concepts, 19; and negative emotions, 170–173; and physical symptoms of emotion, 164–167; and Smell A Memory project, 51–53; and stress and help seeking, 173–175; study of emotion and, 159–162, 167–169; and translation of smells, 48–51
culture cycle, 177
Cultured Pickle Shop, 26–28
current steering technology, 69
Cyberball, 143–145
cyborgs, 72–73, 205–207, 213–215, 218–219, 228, 238, 246–247

dashi, 18–19
da Vinci Surgical System, 94–98, 101, 103–105, 106
decoding, 75, 76, 82
de Heer, Wendy, 74–75, 76–77, 90
Delwiche, Jeannine, 12

De Maria, Walter, 136
dementia. *See* Alzheimer's disease
Dennett, Daniel, 123
depression, 160, 166, 170–173
Deus Ex: Human Revolution, 207
Devanand, Davangere, 46–47
DeWall, C. Nathan, 144–146, 148–149, 153–154
Difference Engine, 134
DiMaio, Simon, 103–104, 106, 107
Dimoveo, Lucas, 238, 239, 241–242, 249
DIYbio ("do-it-yourself biology"), x. *See also* biohacking and biohackers
DNA, and study of taste, 12
Doktor Sleepless (Ellis), x–xi, 238, 253
dopamine, 155–156
dorsomedial prefrontal cortex, 148
Doty, Richard, 44–46, 47–48
dreams, 91–92
drugs, as pain relief, 140, 143, 144–146, 152
Duncan, Jacque, 64, 73
Dzokoto, Vivian, 161

Eagleman, David, 123, 253
earth, movement of, 129
echolocation, 233–234
Einstein, Albert, 131
Eisenberger, Naomi, xvi, 138–142, 147–148, 153–157
electrode electrocorticography (ECOG), 80, 81
electromagnetic induction, 234–235
electrosensors, 234–235
Ellis, Warren, x–xi, 238
emotion(s): categorization of, 163; culture and ideal affect, 169–170; defining, 162–163; language and, 163–164; negative, 170–173; physical feedback and, 164–167; smell and, 37–38; soft biohacking and, xv; study of culture and, 159–162, 167–169; touch and, 221–222; writing in, of others, 175–178
encoding, 75, 82
Eno, Brian, 118, 119
excitement, 176–178
experience, and translation of smells, 48–53
exposure therapy, 184, 188
Eyeborg. *See* Spence, Rob (Eyeborg)

eyecams, 203–204, 205, 207, 208, 220, 228–229
EyeTap, 207

Facial Action Coding System (FACS coding), 161
facial recognition, 211, 215
Farley, Kevin, 26–27
fat, tasting, 3–11
feedback loops, 111, 164–166, 215
fluorescent biomarkers, 106
force feedback, 97, 98–100
Ford, Rob, 219
fountain clocks, 128, 130
Fox, Jesse, 193
frequency, and operation of atomic clock, 129
Friedman, Aaron, 74, 76–77
friendship, as pain relief, 155–157
fusiform gyrus, 178

Gadziola, Marie, 43
Gallant, Jack, 75–76, 84–89, 90–91, 92–93
ganglion cells, 58, 66, 67
Garneau, Nicole, xv, 7–8, 9–11
Geiger, Lazarus, 7
general relativity, 131
gippeun-mat, 22
Glyph, 213
Goel, Anubhuti, 131–133
Gold, Alexandra, 159
Google Glass, 204, 216–217
Gottfried, Jay, 35, 36, 44
Graafstra, Amal, 246, 247, 248, 252
Grammatis, Kosta, 206–207, 208, 228, 229
gravitational redshift, 131
Greenberg, Rob, 61
Greenwich Mean Time, 126–127
"Greenwich pips," 127
grinders, ix, x. *See also* biohacking and biohackers
Grindhouse, ix–xi, 237–242, 247–249, 250–251, 253
group harmony, 170, 173, 174–175
guilt, 183–184
gut feelings, 176, 177

haptic jamming, 100–101, 106
Haraway, Donna, 214

Harbisson, Neil, 207, 216
Haworth, Steve, 242–245
Hayashi, Yuko, 23–24, 25
hearing: and brain activity, 79–81;
    categorical perception and, 78–79;
    and cochlear implants, 60–61; human
    limitations in, 233–234; pain and,
    142; and perception of time, 122–123;
    process of, 77–78; and stimulus
    reconstruction, 81–84; study of brain
    and, 74–77
help seeking, culture and, 173–175
Henderson, Jason, 207
Herz, Rachel, 37, 38, 47
Hillis, Danny, 118
Hodges, Larry, 188
homuncular flexibility, 200–202
homunculus, 200
Hozven, Alex, 26–28
Huffman, Todd, 243, 252–253
Huth, Alex, 74–75, 76–77, 88, 90, 91

ideal affect, 169
identification, and stimulus reconstruction,
    85–86
Ikeda, Kikunae, 4, 18, 19
imagery: auditory, 83; neural activity
    for sensory perception versus, 90;
    reconstruction of, 90–91
Inbar, Ori, 204, 205, 225–226
Industrial Revolution, 126
Innovega, 209–210, 217
insular cortex, 143
Intuitive Surgical, 103–106
iOptik, xix, 209–210
Ivry, Richard, 121

Jarc, Anthony, 103–104, 105, 106
Jarrell, Jesse, 242
John, Bruce, 183, 185–186
Johnsen, Sonke, 234–235, 237, 252

Kamitani, Yukiyasu, 91
Kao, Jonathan, 108–109, 110, 111
Karutz, Cody, 193–195, 197–202
kasu, 26, 27
Keane, Leta, 9
Kelly, Kevin, 228
Kish, Daniel, 233–234

Kissenger, 222–223
Kline, Nathan, 213–214
Knutson, Brian, 175
koji, 23, 24, 25, 26, 27
koji salt, 23, 24
kokumi, 16–18, 20–27
Kross, Ethan, 144

labor, time and, 126
lactisole, 15
language: and description of unestablished
    concepts, xiv–xv, 6–8, 10–11, 13–16,
    22–23; emotion and, 163–164; kokumi
    and problem of, 18; pain and, xvi
Lanier, Jaron, 187, 200
Lao, Alexander, 95
laparoscopic surgery, 102, 105
leadership, and perception of others'
    emotions, 176–178
Lee, Rich, 245–246, 247–248
Lefty O'Doul's, 149–151, 157–158
Licina, Gabriel, 246, 248, 253
Lightning Field, The (De Maria), 136
limbic system, 163
Ling, Martin, 206
linoleic acid, 3–4, 6, 9–10
lipase, 5
Lloyd, Dean, xvii, 54–57, 59–60, 62–65,
    70–73
Long Now Foundation, The, 117–119,
    124–125, 133–137
love: and pain of breakups, 146–147,
    150–152; as pain relief, 155–157
Lowe, John, 128, 129, 130
Ludlow, Andrew, 130, 131
Ly, Daphne, 101, 103
Lyons, Harold, 128
Lyons, Ian, 154

M100 Smart Glasses, 211–212
Mackey, Sean, 142–143, 152–153, 155–156
MacLean, Paul, 163
magnetite hypothesis, 235, 236, 237
magnetoreception: experience of, 242–245,
    252–253; function of, 234–236, 251–252;
    in humans, 230–233, 236–237
"Making the Modern World" collection, 124
Manetta, Celine, 50
Mann, Alfred, 61

Mann, Steve, 207, 212, 216, 219
marijuana, as emotional pain relief,
    145–146
Marsh, Jay, 210, 211, 217
Massenet, Alienor, xv, 48–51
math anxiety, 154–155
Mattes, Richard, 4–5, 9, 10
Mech, Brian, 59, 61, 62, 65–66, 67–68, 69
mechanoreceptors, 98, 109, 142
Meck, Warren, 121, 122
media, 196
medial pre-frontal cortex, 178
Meissner corpuscles, 98
memory: reconstruction of, 91–92; smell
    and, 37–38. *See also* Alzheimer's disease
Merkel cell neurite complexes, 98
Meta, 212
Meta 1, 212
Meta Cookie, 223
Microsoft Kinect, 196
Mirza, Kayvan, 213, 217, 227
miso, 26–27
mixed reality. *See* augmented reality (AR)
Mongelli, Lisa, 149–151
motion, vision and, 70–71
Moulias, Sophie, 30, 31, 34, 47
mouthfeel, 4–5
movement, vision and, 70–71
movie trailers, and stimulus reconstruction,
    87–89
Mugaritz, 222
mutual constitution, 177

Nackley, John, 138, 139–140, 146–147, 149
Naselaris, Thomas, 90
National Institute of Standards and
    Technology (NIST) lab, 127
negative states, 170–172
neural dynamics, time perception through,
    121–123, 131–133
neurofibrillary tangles, 39–41
neuroprosthetics: functioning of, 55, 59;
    touch and, 108–113. *See also* biohacking
    and biohackers; cochlear implants;
    retinal implants
Night Light, 138–140
Nirenberg, Sheila, 66–67, 112
Nishimoto, Shinji, 87, 88
nociceptors, 96, 142

North Paw, 240
Northstar, 240, 250–251, 253

Obama, Barack, 176
Oculus Crescent Bay, 212
Oculus Rift, 196, 212–213
Oculus Touch controllers, 213
OFF-center cells, 58
Okamura, Allison, 96–98, 100, 101,
    106–107, 109
ON-center cells, 58
Opsahl, Kurt, 217–218, 219–220, 227
optical clock, 130–131
optogenetics, 66, 112–113, 132
orrery, 137
Oscar Mayer, 222
O'Shea, Dan, 110–111
Osterhout Design Group, 212
overt speech, 83–84

pacemaker-accumulator model of time
    perception, 120–121
Pacinian corpuscles, 98
pain: brain and emotional, 140–141;
    brain and experience of, 143; defining
    and describing, 141–143; intensity of
    social, 146–147, 151–152; physical and
    emotional, xvi, 143–144; purpose of,
    152–154; recalling, 147–149, 151;
    remedies for, 144–146, 152; research
    on, xvi; as response to perceived threat,
    154–155; and social rejection, 138–140;
    treating, with love or friendship,
    155–157
pain matrix, 143, 155
panic attacks, 164–165
Panksepp, Jaak, 143
Park, BoKyung, 175–176, 178
Pasley, Brian, 81–84, 92
passionate love, 155–156
Pellerin, Laure, 30, 31, 32, 35
perception, xii–xiii; Chang on, 78; cultural,
    of time, 125–127; delay in, 123–124;
    language and, 11; negative emotions
    and, 171–173; of others, 175–178; of
    time, 119, 120–123, 131–133. *See also*
    sensory perception
phobias, virtual reality and treatment of, 188
pixel-perfect registration, 210

plaques, 39–40, 41
polymerase chain reaction (PCR) method, 12
post-traumatic stress disorder (PTSD), 184,
    185–186, 188, 189–191
Poupyrev, Ivan, 223–224, 225
privacy issues: augmented reality and,
    216–218, 219–220; implants and,
    246–247
proprioception, 104, 111
prosthetics, touch and, 108–113
Proteus Effect, 193
Proust, Marcel, 32
PTSD (post-traumatic stress disorder), 184,
    185–186, 188, 189–191

R-7, 212
radical pair hypothesis, 235
railway time, 126
Reed, Danielle, 21–23
relativity, 131
Reppert, Steven, 235–236, 237
Retina Implant AG, 66
retinal implants: development of, 61–62,
    65–67; functioning of, 58–60; future
    possibilities for, 67–70; life and vision
    with, 54–57, 70–73; and relearning
    process, 68
RFID chips, 246, 247
Rhyu, Mee-Ra, 21–22, 23
RingU, 221–222
Ritz, Thorsten, 235
Rizzo, Albert "Skip," 184, 185, 186–188,
    190–191
rods, 57
Rooney, David, 124–127, 133–134
Rose, Alexander, 117–119, 135–137
Rothbaum, Barbara, 188–190
Ruffini endings, 98
Ryder, Andrew, xv, 160, 164–165, 166, 167,
    171, 172–173

saccade movements, 70, 123–124
Saito, Takashi, 24
Salles, Arghavan, 94, 101, 103
Sandoval, Alex, 64
Sarver, Shawn, ix, 238, 239, 241, 251, 253
Scentee, 222
Schorr, Sam, 98–100
Schulten, Klaus, 235

Science for the Masses, 245, 253
Science Museum, 124
Sclafani, Anthony, 16
Second Sight, 55, 61, 65–67, 69–70
self-ownership, 248
sensory perception: biohacking and, 241;
    limitations to, 223, 245–246; neural
    activity for imagery versus, 90
sensory substitution, 96–100, 103, 252–253
Setiyo, Fany, 23, 25
Shenoy, Krishna, xix, 108–110, 112–113
shio koji, 23, 24
Sim City, 187
Sims, Tamara, 177
skeuomorphism, 225
skin stretch, 98–100, 106, 107
smartphones, 197
smell: Alzheimer's disease and loss of,
    38–39, 41–44; and anatomic position of
    olfactory processing, 37; and Atelier
    Olfactif, 29–32; and augmented reality,
    222–225; culture and translation of,
    48–51; evolution of, 36; and flavor
    perception, 8; forgetting and, xv;
    memory and, 37–38; process of, 36–37;
    relationship to memories and emotional
    responses, 32–35; study of, 35; and
    testing for Alzheimer's disease, 44–48
Smell A Memory project, 51–52
smile, 169, 176–178
social anxiety, 166, 173
social norms, and augmented reality,
    226–227
social pain: brain and study of, 140–141;
    intensity of, 146–147, 151–152; purpose
    of, 152–154; recalling, 147–149, 151;
    relationship between physical pain and,
    143–144; remedies for, 144–146, 152
somatosensation, 98, 104, 142
somatosensory cortex, 143
sonar, 233–234
Sood, Raghav, 212
sousveillance, 219–220
space: perception of, 120; time and
    measurement of, 125–126
speech: and brain activity, 79–81;
    categorical perception and, 78–79; and
    stimulus reconstruction, 81–84, 90,
    92–93

Spence, Rob (Eyeborg), xvi–xvii, 203–204, 205–208, 218–219, 220
Stanley, Andrew, 100–101
Starner, Thad, 207
state-dependent learning, 190–191
Stavisky, Sergey, 108–109, 110
Stevenson Won, Andrea, 201, 202
stimulus reconstruction: hearing and, 81–84; obstacles to, 91–93; possibilities and uses for, 89–91, 92–93; purpose of, 75; vision and, 84–89
Stonehenge, 135–136
Stop the Cyborgs, xxi, 215, 216, 247
stress, culture and help seeking and, 173–175
striatal-beat frequency model of time perception, 121
STRIVE (Stress Resilience in Virtual Environments) program, 185, 190–191
superior temporal gyrus, 79–80, 81–82, 83
surgery: with da Vinci Surgical System, 94–98, 101; development of robotic, 105–107; long-distance, 107–108; vision and touch in, 102–103; working by hand in, 102
surveillance: augmented reality and, 216–218, 219–220; implants and, 246–247
sweet receptor, xiii–xiv, 11–12

Tag-Candy, 223
taijin kyofusho, 173, 175
Tan, Ai-lin, 51, 52, 53
tangles, 39–41
taste: in animals, xiii–xiv; and augmented reality, 222–225; categorical perception and, 79; and change preceding acceptance of umami, 18–21; describing sixth, xiv–xv; of fat, 3–11; kokumi and, 16–18, 21–27; and lexical challenge of describing unestablished concepts, 11–16; study of pain and, 139
tau, 40
teleoperation, 105, 107–108
telepresence, 105
telesurgery, 105, 107–108
10,000 Year Clock, 117–119, 124–125, 133–137
Terho, Miikka, 207
Tewell, Jordan, 221

thermoreceptors, 98
time: and atomic clock, 127–131; cultural perception of, 125–127; as cultural phenomenon, 119–120; perception of, 119, 120–123, 131–133; and perceptual delay, 123–124
Tordoff, Michael: on basic taste, 13; on differences in taste, xiii–xiv; on lexical challenge of describing unestablished concepts, 11, 14–16; on studying internal processes, xviii
touch: and augmented reality, 221–225; and development of robotic surgery, 105–107; and haptic jamming, 100–101; and magnetoreception, 252–253; neuroprosthetics and, 108–113; pain and, 142; and sensory substitution, 96–100; vision and, 102–103, 106
transduction, 55
"transfer" (shift in feeling), 193–194
travel, time and coordination of, 126
Tsai, Jeanne, 160, 161, 162–163, 169–170, 176–178
Tylenol, as emotional pain relief, 144–145

umami, 4, 12, 18–21
uncertainty, and pain of social rejection, 146–147
UPSIT (University of Pennsylvania Smell Identification Test), 45–47, 49, 50

valence, and categorization of emotion, 163
Valentin, Dominique, 50
Vaterlaus, Mandi, 244
ventral medial prefrontal cortex, 156–157
VeriChip, 246
video, and stimulus reconstruction, 87–89
Virtual Human Interaction Lab, 193–202
Virtual Iraq, 190
virtual reality (VR), xix; evolution of, 186–188; experience of, 191–192, 197–200; and homuncular flexibility, 200–202; and influencing behavior, 192–199; military use of, 183–186; and treatment of phobias and PTSD, 188–191
Virtual Vietnam, 189–190
visceral brain, 163

vision: and development of retinal implants,
    61–62, 65–67; and future possibilities
    for retinal implants, 67–70; human
    limitations in, 233; Lloyd's loss of,
    62–64; pain and, 142; process of, 57–58;
    with retinal implant, 54–57, 58–60,
    70–73; retinal implants and relearning,
    68; and stimulus reconstruction, 84–91;
    touch and, 102–103, 106. *See also*
    augmented reality (AR)
Vlach, Tanya Marie, 227–229
Von Cyborg, Aneta, 231
Von Cyborg, Samppa, 231–232, 250
Vuzix, 211–212

Warwick, Kevin, 214, 246
Watanabe, Momoka, 173, 174

Wesson, Daniel, 39–40, 41–44, 251–252
Williams, Kipling, 143, 148
Wolbring, Gregor, 226, 247, 249
Wood, Adam, 215–216, 219, 220,
    227, 247
Wren, Sherry, xviii, 94–96, 101–103, 106,
    107, 201

Yee, Nick, 193
Yoshida, Shintaro, 17, 19–20
Younger, Jarred, 155
ytterbium optical lattice clock, 130–131
Yuzuki Japanese Eatery, 23–26

Zhang, Arthur, 209, 210, 217
Zhao, Yue, 173, 174
Zhou, Biru, 173–175

Credit: © Justine Quart

KARA PLATONI has a master's degree from UC Berkeley's Graduate School of Journalism, where she now teaches reporting and narrative writing. The winner of the American Association for the Advancement of Science's Journalism Award and the Evert Clark/Seth Payne Award for Young Science Journalists, she lives in Oakland, California.